Wilkie Collins

Rambles Beyond Railways

Notes Taken in Cornwell

Wilkie Collins

Rambles Beyond Railways
Notes Taken in Cornwell

ISBN/EAN: 9783744678537

Printed in Europe, USA, Canada, Australia, Japan

Cover: Foto ©berggeist007 / pixelio.de

More available books at **www.hansebooks.com**

LAMORNA COVE.

RAMBLES BEYOND RAILWAYS;

OR,

Notes in Cornwall taken A-Foot.

By WILKIE COLLINS,
AUTHOR OF
"ANTONINA," "THE WOMAN IN WHITE," ETC.

The Land's End, Cornwall.

NEW EDITION.

LONDON:
RICHARD BENTLEY: NEW BURLINGTON STREET.
Publisher in Ordinary to Her Majesty.
1865.

DEDICATED TO

THE COMPANION OF MY WALK THROUGH CORNWALL,

HENRY C. BRANDLING.

PREFACE

TO

THE PRESENT EDITION.

I visited Cornwall, for the first time, in the summer and autumn of 1850; and in the winter of the same year, I wrote this book.

At that time, the title attached to these pages was strictly descriptive of the state of the county, when my companion and I walked through it. But when, little more than a year afterwards, a second edition of this volume was called for, the all-conquering railway had invaded Cornwall in the interval, and had practically contradicted me on my own title-page.

To rechristen my work was out of the question—I should simply have destroyed its individuality. Ladies may, and do, often change their names for the better; but books enjoy no such privilege. In this embarrassing position, I ended by treating the ill-timed intrusion of the railway into my literary affairs, as a certain Abbé (who was also an author,) once treated the overthrow of the Swedish Constitution, in the reign of Gustavus the Third. Having written a profound work, to prove that the Constitution, as at that time settled, was secure from all political accidents, the Abbé was surprised in his study, one day, by the appearance of a gentleman, who disturbed him over the correction of his last proof-sheet. "Sir!" said the gentleman; "I have looked in to inform you that the Constitution has just been overthrown." To which the Abbé replied:— "Sir! they may overthrow the Constitution, but they can't overthrow MY BOOK"—and he quietly went on with his work.

On precisely similar principles, I quietly went on with MY TITLE-PAGE.

So much for the name of the book. For the book itself, as published in its present form, I have a last word to say, before these prefatory lines come to an end.

Cornwall no longer offers the same comparatively untrodden road to the literary traveller which it presented when I went there. Many writers have made the journey successfully, since my time. Mr. Walter White, in his "Londoner's Walk to the Land's End," has followed me, and rivalled me, on my own ground. Mr. Murray has published "The Handbook to Cornwall and Devon"—and detached essays on Cornish subjects, too numerous to reckon up, have appeared in various periodical forms. Under this change of circumstances, it is not the least of the debts which I owe to the encouraging kindness of my readers, that they have not forgotten "Rambles Beyond Railways," and that the continued demand for the book is such as to justify the appearance of the present edition. I have, as I believe, to thank the unambitious purpose with which I originally wrote, for thus keeping me in remembrance. All that my book attempts is frankly to record a series of personal impressions; and, as a necessary consequence—though my title is obsolete, and my pedestrian adventures are old-fashioned—I have a character of my own still left, which readers can recognise; and the homely travelling narrative which I brought from Cornwall, eleven years since, is not laid on the shelf yet.

I have spared no pains to make these pages worthy of the approval of new readers. The book has been carefully revised throughout; and certain hastily-written passages, which my better experience condemns as unsuited to the main design, have been removed altogether. Two of the lithographic illustrations, (now no longer in existence) with which my friend and fellow-traveller, Mr. Brandling, adorned the previous editions, have been copied on wood, as accurately as circumstances would permit; and a "Postscript" has been added, which now appears in connexion with the original narrative, for the first time.

The little supplementary sketch thus presented, describes a cruise to the Scilly Islands, (taken five years after the period of my visit to Cornwall), and completes the round of my travelling experiences in the far West of England. These newly-added pages are written, I am afraid, in a tone of somewhat boisterous gaiety—which I have not, however, had the heart to subdue, because it is after all the genuine offspring of the "harum-scarum" high spirits of the time. The "Cruise of the Tomtit" was, from first to last, a practical burlesque; and the good-natured reader will, I hope, not think the worse of me, if I beg him to stand on no ceremony, and to laugh his way through it as heartily as he can.

HARLEY STREET, LONDON,
March, 1861.

CONTENTS.

	PAGE
I. A Letter of Introduction	1
II. A Cornish Fishing Town	5
III. Holy Wells and Druid Relics	23
IV. Cornish People	55
V. Loo-pool	86
VI. The Lizard	97
VII. The Pilchard Fishery	120
VIII. The Land's End	139
IX. Botallack Mine	155
X. The Modern Drama in Cornwall	180
XI. The Ancient Drama in Cornwall	197
XII. The Nuns of Mawgan	216
XIII. Legends of the Northern Coast	231

POSTSCRIPT.

The Cruise of the Tomtit to the Scilly Islands	253

RAMBLES BEYOND RAILWAYS.

I.

A LETTER OF INTRODUCTION.

Dear Reader,

When any friend of yours or mine, in whose fortunes we take an interest, is about to start on his travels, we smooth his way for him as well as we can, by giving him a letter of introduction to such connexions of ours as he may find on his line of route. We bespeak their favourable consideration for him by setting forth his good qualities in the best light possible; and then leave him to make his own way by his own merit — satisfied that we have done enough in procuring him a welcome under our friend's roof, and giving him at the outset a claim to our friend's estimation.

Will you allow me, reader (if our previous

acquaintance authorizes me to take such a liberty), to follow the custom to which I have just adverted; and to introduce to your notice this Book, as a friend of mine setting forth on his travels, in whose well-being I feel a very lively interest. He is neither so bulky nor so distinguished a person as some of the predecessors of his race, who may have sought your attention in years gone by, under the name of "Quarto," and in magnificent clothing of Morocco and Gold. All that I can say for his outside is, that I have made it as neat as I can—having had him properly thumped into wearing his present coat of decent cloth, by the most competent book-tailor I could find. As for his intrinsic claims to your kindness, he has only two that I shall venture to advocate. In the first place he is able to tell you something about a part of your own country which is still too rarely visited and too little known. He will speak to you of one of the remotest and most interesting corners of our old English soil. He will tell you of the grand and varied scenery; the mighty Druid relics; the quaint legends; the deep, dark mines; the venerable remains of early Christianity; and the pleasant primitive population of the county of CORNWALL. You will inquire, can we believe him in all that he says? This brings me at once to his second qualification—he invariably speaks the truth. If he describes scenery to you, it is scenery that he saw and noted on the spot; and if he adds some little sketches of character, I answer for him, on my own responsibility, that they are sketches drawn from the life.

Have I said enough about my friend to interest you in his fortunes, when you meet him wandering hither and thither over the great domain of the Republic of Letters—or, must I plead more warmly in his behalf? I can only urge on you that he does not present himself as fit for the top seats at the library table,—as aspiring to the company of those above him,—of classical, statistical, political, philosophical, historical, or antiquarian high dignitaries of his class, of whom he is at best but the poor relation. Treat him not, as you treat such illustrious guests as these! Toss him about anywhere, from hand to hand, as goodnaturedly as you can; stuff him into your pocket when you get into the railway; take him to bed with you, and poke him under the pillow; present him to the rising generation, to try if he can amuse *them;* give him to the young ladies, who are always predisposed to the kind side, and may make something of him; introduce him to "my young masters" when they are idling away a dull morning over their cigars. Nay, advance him if you will, to the notice of the elders themselves; but take care to ascertain first that they are people who only travel to gratify a hearty admiration of the wonderful works of Nature, and to learn to love their neighbour better by seeking him at his own home—regarding it, at the same time, as a peculiar privilege, to derive their satisfaction and gain their improvement from experiences on English ground. Take care of this; and who knows into what high society you may not be able to introduce the bearer of the present letter! In spite of his habit of rambling from subject to subject in his talk, much as he rambled from place to place in his travels, he may actually find himself, one day, basking on Folio Classics beneath the genial approval of a Doctor of Divinity, or trembling among Statutes and Reports under the learned scrutiny of a Sergeant at Law!

W. C.

Harley Street, London,
March, 1861.

II.

A CORNISH FISHING TOWN.

The time is ten o'clock at night—the scene, a bank by the road-side, crested with young fir-trees, and affording a temporary place of repose to two travellers, who are enjoying the cool night air, picturesquely extended flat on their backs—or rather, on their knapsacks, which now form part and parcel of their backs. These two travellers are, the writer of this book, and an artist friend who is the companion of his rambles. They have long desired to explore Cornwall together, on foot; and the object of their aspirations has been at last accomplished, in the summer-time of the year eighteen hundred and fifty.

In their present position, the travellers are (to speak geographically) bounded towards the east by a long road winding down the side of a rocky hill;

towards the west, by the broad half-dry channel of a tidal river; towards the north, by trees, hills, and upland valleys; and towards the south, by an old bridge and some houses near it, with lights in their windows faintly reflected in shallow water. In plainer words, the southern boundary of the prospect around them represents a place called Looe—a fishing-town on the south coast of Cornwall, which is their destination for the night.

They had, by this time, accomplished their initiation into the process of walking under a knapsack, with the most complete and encouraging success. You, who in these days of vehement bustle, business, and competition, can still find time to travel for pleasure alone—you, who have yet to become emancipated from the thraldom of railways, carriages, and saddle-horses—patronize, I exhort you, that first and oldest-established of all conveyances, your own legs! Think on your tender partings nipped in the bud by the railway bell; think of crabbed cross-roads, and broken carriage-springs; think of luggage confided to extortionate porters, of horses casting shoes and catching colds, of cramped legs and numbed feet, of vain longings to get down for a moment here, and to delay for a pleasant half hour there—think of all these manifold hardships of riding at your ease; and the next time you leave home, strap your luggage on your shoulders, take your stick in your hand, set forth delivered from a perfect paraphernalia of incumbrances, to go where you will, how you will—the free citizen of the whole travelling world! Thus independent, what may you not accomplish?—what pleasure is there that you cannot enjoy? Are you an artist?—you can stop to sketch every point of view that strikes your eye. Are you a philanthropist?—you can go into every cottage and talk to every human being you pass. Are you a botanist, or geologist?—you may pick up leaves and chip rocks wherever you please, the live-long day. Are you a valetudinarian?—you may physic yourself by Nature's own simple prescription, walking in fresh air. Are you dilatory and irresolute?—you may dawdle to your heart's content; you may change all your plans a dozen times in a dozen hours; you may tell "Boots" at the inn to call you at six o'clock, may fall asleep again (ecstatic sensation!) five minutes after he has knocked at the door, and may get up two hours later, to pursue your journey, with perfect impunity and satisfaction. For, to you, what is a time-table but waste-paper?—and a "booked place" but a relic of the dark ages? You dread, perhaps, blisters on your feet—sponge your feet with cold vinegar and water, change your socks every ten miles, and show me blisters after that, if you can! You strap on your knapsack for the first time, and five minutes afterwards feel an aching pain in the muscles at the back of your neck—walk *on*, and the aching will walk *off!* How do we overcome our first painful cuticular reminiscences of first getting on horseback?—by riding again. Apply the same rule to carrying the knapsack, and be assured of the same successful result. Again I say it, therefore—walk, and be merry; walk, and be healthy; walk, and be your own master!—walk, to enjoy, to observe, to improve, as no riders can!—walk, and you are the best peripatetic impersonation of holiday enjoyment that is to be met with on the surface of this work-a-day world!

How much more could I not say in praise of travelling on our own neglected legs? But it is getting late; dark night-clouds are marching slowly over the sky, to the whistling music of the wind; we must leave our bank by the road-side, pass one end of the old bridge, walk along a narrow winding street, and enter our hospitable little inn, where we are welcomed by the kindest of landladies, and waited on by the fairest of chambermaids. If Looe prove not to be a little sea-shore paradise to-morrow, then is there no virtue in the good omens of to-night.

* * * * *

The first point for which we made in the morning, was the old bridge; and a most picturesque and singular structure we found it to be. Its construction dates back as far as the beginning of the fifteenth century. It is three hundred and eighty-four feet long, and has fourteen arches, no two of which are on the same scale. The stout buttresses built between each arch, are hollowed at the top into curious triangular places of refuge for pedestrians, the roughly paved roadway being just wide enough to allow the passage of one cart at a time. On some of these buttresses, towards the middle, once stood an oratory, or chapel, dedicated to St. Anne; but no vestiges of it now remain. The old bridge however, still rises sturdily enough on its ancient foundations; and, whatever the point from which its silver-grey stones and quaint arches of all shapes and sizes may be beheld, forms no mean adjunct to the charming landscape around it.

Looe is known to have existed as a town in the reign of Edward I.; and it remains to this day one of the prettiest and most primitive places in England. The river divides it into East and West Looe; and the view from the bridge, looking towards the two little colonies of houses thus separated, is in some respects almost unique.

At each side of you rise high ranges of beautifully wooded hills; here and there a cottage peeps out among the trees, the winding path that leads to it being now lost to sight in the thick foliage, now visible again as a thin serpentine line of soft grey. Midway on the slopes appear the gardens of Looe, built up the acclivity on stone terraces one above another; thus displaying the veritable garden architecture of the mountains of Palestine magically transplanted to the side of an English hill. Here, in this soft and genial atmosphere, the hydrangea is a common flower-bed ornament, the fuchsia grows lofty and luxuriant in the poorest cottage garden, the myrtle flourishes close to the sea-shore, and the tender tamarisk is the wild plant of every farmer's hedge. Looking lower down the hills yet, you see the houses of the town straggling out towards the sea along each bank of the river, in mazes of little narrow streets; curious old quays project over the water at different points; coast-trade vessels are being loaded and unloaded, built in one place and repaired in another, all within view; while the prospect of hills, harbour, and houses thus quaintly combined together, is beautifully closed by the English Channel, just visible as a small strip of blue water, pent in between the ridges of two promontories which stretch out on either side to the beach.

Such is Looe as beheld from a distance; and it loses none of its attractions when you look at it more closely. There is no such thing as a straight street in the place. No martinet of an architect has been here, to drill the old stone houses into regimental regularity. Sometimes you go down steps into the ground floor, sometimes you mount an outside staircase to get to the bed-rooms. Never were such places devised for hide and seek since that exciting nursery pastime was first invented. No house has fewer than two doors leading into two different lanes; some have three, opening at once into a court, a street, and a wharf, all situated at different points of the compass. The shops, too, have their diverting irregularities, as well as the town. Here you might call a man a Jack of all trades, as the best and truest compliment you could pay him—for here one shop combines in itself a drug-mongering, cheese-mongering, stationery, grocery, and oil and Italian line of business; to say nothing of such cosmopolitan miscellanies as wrinkled apples, dusty nuts, cracked slate pencils and fly-blown mock jewellery. The moral good which you derive, in the first pane of a window, from the contemplation of memoirs of murdered missionaries and serious tracts against intemperance and tight-lacing, you lose in the second, before such worldly temptations as gingerbread, shirt-studs, and fascinating white hats for Sunday wear, at two and ninepence a piece. Let no man rashly say he has seen all that British enterprise can do for the extension of British commerce, until he has carefully studied the shop-fronts of the tradesmen of Looe.

Then, when you have at last threaded your way successfully through the streets, and have got out on the beach, you see a pretty miniature bay, formed by the extremity of a green hill on the right, and by fine jagged slate-rocks on the left. Before this seaward quarter of the town is erected a strong bulwark of rough stones, to resist the incursion of high tides. Here, the idlers of the place assemble to lounge and gossip, to look out for any outward-bound ships that are to be seen in the Channel, and to criticise the appearance and glorify the capabilities of the little fleet of Looe fishing-boats, riding snugly at anchor before them at the entrance of the bay.

The inhabitants number some fourteen hundred; and are as good-humoured and unsophisticated a set of people as you will meet with anywhere. The Fisheries and the Coast Trade form their principal means of subsistence. The women take a very fair share of the hard work out of the men's hands. You constantly see them carrying coals from the vessels to the quay in curious hand-barrows: they laugh, scream, and run in each other's way incessantly: but these little irregularities seem to assist, rather than impede them, in the prosecution of their tasks. As to the men, one absorbing interest appears to govern them all. The whole day long they are mending boats, painting boats, cleaning boats, rowing boats, or, standing with their hands in their pockets, looking at boats. The children seem to be children in size, and children in nothing else. They congregate together in sober little groups, and hold mysterious conversations, in a dialect which we cannot understand. If they ever do tumble down, soil their pinafores, throw stones, or make mud pies, they practise these juvenile vices in a midnight secrecy which no stranger's eye can penetrate.

In that second period of the dark ages, when there were High Tories and rotten boroughs in the land, Looe (containing at that time nothing like the number of inhabitants which it now possesses) sent Four Members to Parliament! The ceremony by which two of these members were elected, as it was described to me by a man who remembered witnessing it, must have been an impressive sight indeed to any foreigner interested in studying the representative system of this country. On the morning of the "Poll," one division of the borough sent *six* electors, and another *four*, to record their imposing aggregate of votes in favour of any two smiling civil gentlemen, who came, properly recommended, to ask for them. This done, the ten electors walked quietly home in one direction, and the two members walked quietly off in another, to perform the fatiguing duty of representing their constituents' interests in Imperial Parliament. The election was quite a snug little family affair, in these "good old times." The ten gentlemen who voted, and the other two gentlemen who took their votes, just made up a comfortable compact dozen, all together!

But this state of things was too harmonious to last in such a world of discord as ours. The day of innovation came: turbulent Whigs and Radicals laid uncivil hands on the Looe polling-booth, and politically annihilated the pleasant party of twelve. Since that disastrous period the town has sent no members to Parliament at all; and very little, indeed, do the townspeople appear to care about so serious a deprivation. In case the reader should be disposed to attribute this indifference to municipal privileges to the supineness rather than the philosophy of the inhabitants, I think it necessary to establish their just claims to be considered as possessing public spirit, prompt decision, and wise fertility of resource in cases of emergency, by relating in this place the true story of how the people of Looe got rid of the rats.

About a mile out at sea, to the southward of the town, rises a green triangular shaped eminence, called Looe Island. Here, many years ago, a ship was wrecked. Not only were the sailors saved, but several free passengers of the rat species, who had got on board, nobody knew how, where, or when, were also preserved by their own strenuous exertions, and wisely took up permanent quarters for the future on the terra firma of Looe Island. In process of time, and in obedience to the laws of nature, these rats increased and multiplied exceedingly; and, being confined all round within certain limits by the sea, soon became a palpable and dangerous nuisance. Destruction was threatened to the agricultural produce of all the small patches of cultivated land on the island—it seemed doubtful whether any man who ventured there by himself, might not share the fate of Bishop Hatto, and be devoured by rats. Under these pressing circumstances, the people of Looe determined to make one united and vehement effort to extirpate the whole colony of invaders. Ordinary means of destruction had been tried already, and without effect. It was said that rats left for dead on the ground had mysteriously revived faster than they could be picked up and skinned, or flung into the sea. Rats desperately wounded had got away into their holes, and become convalescent, and increased and multiplied again more productively than ever. The great problem was, not how to kill the rats, but how to annihilate them so effectually as to place the re-appearance even of one of them altogether out of the question. This was the problem, and it was solved in the following manner:—

All the available inhabitants of the town were called to join in a great hunt. The rats were caught by every conceivable artifice; and, once taken, were instantly and ferociously *smothered in onions:* the corpses were then decently laid out on clean china dishes, and straightway eaten with vindictive relish by the people of Looe. Never was any invention for destroying rats so complete and so successful as this! Every man, woman, and child, who could eat, could swear to the extirpation of all the rats they had eaten. The local returns of dead rats were not made by the bills of mortality, but by the bills of fare: it was getting rid of a nuisance by the unheard-of process of stomaching a nuisance! Day after day passed on, and rats disappeared by hundreds, never to return. What could all their cunning and resolution avail them now? They had resisted before, and could have resisted still, the ordinary force of dogs, ferrets, traps, sticks, stones, and guns, arrayed against them; but when to these engines of assault were added, as auxiliaries, smothering onions, scalding stew-pans, hungry mouths, sharp teeth, good digestions, and the gastric juice, what could they do but give in? Swift and sure was the destruction that now overwhelmed them—everybody who wanted a dinner had a strong personal interest in hunting them down to the very last. In a short space of time the island was cleared of the usurpers. Cheeses remained entire: ricks rose uninjured. And this is the true story of how the people of Looe got rid of the rats!

It will not much surprise any reader who has been goodnatured enough to peruse the preceding pages with some attention, to hear that we idly delayed the day of departure from the pleasant fishing-town on the south coast, which was now the place of our sojourn. The smiles of our fair chambermaid and the cookery of our excellent hostess, addressed us in Siren tones of allurement which we had not the virtue to resist. Then, it was difficult to leave unexplored any of the numerous walks in the neighbourhood—all delightfully varied in character, and each possessing its own attractive point of view. Even when we had made our determination and fixed our farewell day, a great boat-race and a great tea-drinking, which everybody declared was something that everybody else ought to see, interfered to detain us. We delayed yet once more, to partake in the festivities, and found that they supplied us with all the necessary resolution to quit Looe which we had hitherto wanted. We had remained to take part in a social failure on a very large scale.

As, in addition to the boat-race, there was to be a bazaar on the beach; and as fine weather was therefore an essential requisite on the occasion, it is scarcely necessary to premise that we had an unusually large quantity of rain. In the forenoon, however, the sun shone with treacherous brilliancy; and all the women in the neighbourhood fluttered out in his beams, gay as butterflies. What dazzling gowns, what flaring parasols, what joyous cavalcades on cart-horses, did we see on the road that led to the town! What a mixture of excitement, confusion, anxiety, and importance, possessed everybody! What frolic and felicity attended the popular gatherings on the beach, until the fatal moment when the gun fired for the first race! Then, as if at that signal, the clouds began to muster in ominous blackness; the deceitful sunlight disappeared; the rain came down for the day—a steady, noiseless, malicious rain, that at once forbade all hope of clear weather. Dire was the discomfiture of the poor ladies of Looe. They ran hither and thither for shelter, in lank wet muslin and under dripping parasols, displaying, in the lamentable emergency of the moment, all sorts of interior contrivances for expanding around them the exterior magnificence of their gowns, which we never ought to have seen. Deserted were the stalls of the bazaar for the parlours of the alehouses; unapplauded and unobserved, strained at the oar the stout rowers in the boat-race. Everybody ran to cover, except some seafaring men who cared nothing for weather, some inveterate loungers who would wander up and down in spite of the rain, and three unhappy German musicians, who had been caught on their travels, and pinned up tight against the outer wall of a house, in a sort of cage of canvas, boards, and evergreens, which hid every part of them but their heads and shoulders. Nobody interfered to release these unfortunates. There they sat, hemmed in all round by dripping leaves, blowing grimly and incessantly through instruments of brass. If the reader can imagine the effect of three phlegmatic men with long bottle noses, looking out of a circle of green bushes, and playing waltzes unintermittingly on long horns, in a heavy shower—he will be able to form a tolerably correct estimate of the large extra proportion of gloom which the German musicians succeeded in infusing into the disastrous proceedings of the day.

The tea-drinking was rather more successful. The room in which it was held was filled to the corners, and exhaled such an odour of wet garments and bread and butter (to say nothing of an incessant clatter of china and bawling of voices) that we found ourselves, as uninitiated strangers, unequal to the task of remaining in it to witness the proceedings. Descending the steps which led into the street from the door—to the great confusion of a string of smartly dressed ladies who encountered us, rushing up with steaming tea-kettles and craggy lumps of plumcake—we left the inhabitants to conclude their festivities by themselves, and went out to take a farewell walk on the cliffs of Looe.

We ascended the heights to the westward, losing sight of the town among the trees as we went; and then, walking in a southerly direction through some cornfields, approached within a few hundred yards of the edge of the cliffs, and looked out on the sea. The sky had partially cleared, and the rain had ceased; but huge fantastic masses of cloud, tinged with lurid copper-colour by the setting sun, still towered afar off over the horizon, and were reflected in a deeper hue on the calm surface of the sea, with a perfectness and grandeur that I never remember to have witnessed before. Not a ship was in sight; but out on the extreme line of the wilderness of grey waters there shone one red, fiery spark—the beacon of the Eddystone Lighthouse. Before us, the green fields of Looe Island rose high out of the ocean—here, partaking the red light on the clouds; there, half lost in cold shadow.

Closer yet, on the mainland, a few cattle were feeding quietly on a long strip of meadow bordering the edge of the cliff; and, now and then, a gull soared up from the sea, and wheeled screaming over our heads. The faint sound of the small shore-waves (invisible to us in the position we occupied) beating dull and at long intervals on the beach, augmented the dreary solemnity of the evening prospect. Light, shade, and colour, were all before us, arranged in the grandest combinations, and expressed by the simplest forms. If Michael Angelo had painted landscape, he might have represented such a scene as we now beheld.

This was our last excursion at Looe. The next morning we were again on the road, walking inland on our way to the town of Liskeard.

III.

HOLY WELLS AND DRUID RELICS.

FRESH from the quaint old houses, the delightfully irregular streets, and the fragrant terrace-gardens of Looe, we found ourselves, on entering Liskeard, suddenly introduced to that "abomination of desolation," a large agricultural country town. Modern square houses, barren of all outer ornament; wide, dusty, deserted streets; misanthropical-looking shopkeepers, clad in rusty black, standing at their doors to gaze on the solitude around them—greeted our eyes on all sides. Such samples of the population as we accidentally encountered were not promising. We were unlucky enough to remark, in the course of two streets, a nonagenarian old woman with a false nose, and an idiot shaking with the palsy.

But harder trials were in reserve for us. We missed the best of the many inns at Liskeard, and went to the very worst. What a place was our house of public entertainment for a great sinner to repent in, or for a melancholy recluse to retreat to! Not a human being appeared in the street where this tavern of despair frowned amid congenial desolation. Nobody welcomed us at the door—the sign creaked dolefully, as the wind swung it on its rusty hinges. We walked in, and discovered a low-spirited little man sitting at an empty "bar," and hiding himself, as it were, from all mortal inspection behind the full sheet of a dirty provincial newspaper. Doleful was our petition to this secluded publican for shelter and food; and doubly doleful was his answer to our appeal. Beds he believed he had—food there was none in the house, saving a piece of *corned beef*, which the family had dined on, and which he proposed that we should partake of before it got quite cold. Having said thus much, he suddenly retired behind his newspaper, and spoke no word more.

In a few minutes the landlady appeared, looking very thin and care-worn, and clad in mourning weeds. She smiled sadly upon us; and desired to know how we liked corned beef? We acknowledged a preference for fresh meat, especially in large market towns like Liskeard, where butchers' shops abounded. The landlady was willing to see what she could get; and in the meantime, begged to be allowed to show us into a private room. She succeeded in incarcerating us in the most thoroughly private room that could be found out of a model prison. It was situated far away at the back of the house, and looked out upon a very small yard entirely circumscribed by empty stables. The one little window was shut down tight, and we were desired not to open it, for fear of a smell from these stables. The ornaments of the place consisted of hymn-books, spelling-books, and a china statue of Napoleon in a light green waistcoat and a sky-blue coat. There was not even a fly in the room to intrude on us in our privacy; there were no cocks and hens in the yard to cackle on us in our privacy; nobody walked past the outer passage, or made any noise in any part of the house, to startle us in our privacy; and a steady rain was falling propitiously to keep us in our privacy. We dined in our retired situation on some rugged lumps of broiled flesh, which the landlady called chops, and the servant steaks. We broke out of prison after dinner, and roamed the streets. We returned to solitary confinement in the evening, and were instantly conducted to another cell.

This second private apartment appeared to be about forty feet long; six immense wooden tables, painted of a ghastly yellow colour, were ranged down it side by side. Nothing was placed on any of them—they looked like dissecting-tables waiting for "subjects." There was yet another and a seventh table—a round one, half lost in a corner, to which we retreated for refuge—it was covered with crape and bombazine, half made up into mourning garments proper to the first and intensest stage of grief. The servant brought us one small candle to cheer the scene; and desired to be informed whether we wanted *two* sheets apiece to our beds, or whether we could do with a sheet at top and a blanket at bottom, as other people did? This question cowed us at once into gloomy submission to our fate. We just hinted that we had contracted bad habits of sleeping between two sheets, and left the rest to chance; reckless how we slept, or where we slept, whether we passed the night on the top of one of the six dissecting-tables, or with a blanket at bottom, as other people passed it. Soon the servant returned to tell us that we had got our two sheets each, and to send us to bed—snatching up the landlady's mourning garments, while she spoke, with a scared, suspicious look, as if she thought that the next outrageous luxury we should require would be a nightgown apiece of crape and bombazine.

Reflecting on our lamentable situation the last thing at night, we derived some consolation from remembering that we should leave our quarters early the next morning. It was not Liskeard that we had come to see, but the country around Liskeard—the famous curiosities of Nature and Art that are to be found some six or eight miles away from the town. Accordingly, we were astir betimes on the morrow. The sky was fair; the breeze was exhilarating. Once past the doleful doorway of the inn, we found ourselves departing under the fairest auspices for a pilgrimage to the ruins of St. Cleer's Well, and to the granite piles and Druid remains, now entitled the "Cheese-Wring" and "Hurler" rocks.

On leaving the town, our way lay to the northward, up rising ground. For the first two miles, the scenery differed little from what we had already beheld in Cornwall. The lanes were still sunk down between high banks, like dry ditches; all varieties of ferns grew in exquisite beauty and luxuriance on either side of us; the trees were small in size, and thickly clothed with leaves; and the views were generally narrowed to a few well-cultivated fields, with sturdy little granite-built cottages now and then rising beyond. It was only when we had reached what must have been a considerable elevation, that any change appeared in the face of the country. Five minutes more of walking, and a single turn in the road, brought us suddenly to the limits of trees, meadows, and cottages; and displayed before us, with almost startling abruptness, the magnificent prospect of a Cornish Moor.

The expanse of open plain that we now beheld stretched away uninterruptedly on the right hand, as far as the distant hills. Towards the left, the view was broken and varied by some rough stone walls, a narrow road, and a dip in the earth beyond. Wherever we looked, far or near, we saw masses of granite of all shapes and sizes, heaped irregularly on the ground among dark clusters of heath. An old furze-cutter was the only human figure that appeared on the desolate scene. Approaching him to ask our way to St. Cleer's Well—no signs of which could be discerned on the wilderness before us—we found the old fellow, though he was eighty years of age, working away with all the vigour of youth. On this wild moor he had lived and laboured from childhood; and he began to talk proudly of its great length and breadth, and of the wonderful sights that were to be seen on different parts of it, the moment we addressed him. He described to us, in his own homely forcible way, the awful storms that he had beheld, the fearful rattling and roaring of thunder over the great unsheltered plain before us—the hail and sleet driven so fiercely before the hurricane, that a man was half-blinded if he turned his face towards it for a moment—the forked lightning shooting from pitch-dark clouds, leaping and running fearfully over the level ground, blackening, splitting, tearing from their places the stoutest rocks on the moor. Three masses of granite lay heaped together near the spot where we had halted—the furze-cutter pointed to them with his bill-hook, and told us that what we now looked on was once one great rock, which he had seen riven in an instant by the lightning into the fragmentary form that it now presented. If we mounted the highest of these three masses, he declared that we might find out our own way to St. Cleer's Well by merely looking around us. We followed his directions. Towards the east, far away over the magnificent sweep of moorland, and on the slope of the hill that bounded it, appeared the tall chimneys and engine-houses of the Great Caraton Copper Mine—the only objects raised by the hand of man that were to be seen on this part of the view. Towards the west, much nearer at hand, four grey turrets were just visible beyond some rising ground. These turrets belonged to the tower of St. Cleer's Church, and the Well was close by it.

Taking leave of the furze-cutter, we followed the path at once that led to St. Cleer's. Half an hour's walking brought us to the village, a straggling, picturesque place, hidden in so deep a hollow as to be quite invisible from any distance. All the little cottage-girls whom we met, carrying their jugs and pitchers of water, curtseyed and wished us good morning with the prettiest air of bashfulness and good humour imaginable. One of them, a rosy, beautiful child, who proudly informed us that she was six years old, put down her jug at a cottage-gate and ran on before to show us the way, delighted to be singled out from her companions for so important an office. We passed the grey walls of the old church, walked down a lane, and soon came in sight of the Well, the position of which was marked by a ruined Oratory, situated on some open ground close at the side of the public pathway.

St. Cleer, or—as the name is generally spelt out of Cornwall—St. Clare, the patron saint of the Well, was born in Italy, in the twelfth century—and born to a fair heritage of this world's honours and this world's possessions. But she voluntarily abandoned, at an early age, all that was alluring in the earthly career awaiting her; to devote herself entirely to the interests of her religion and the service of Heaven. She was the first woman who sat at the feet of St. Francis as his disciple, who humbly practised the self-mortification, and resolutely performed the vow of perpetual poverty, which her preceptor's harshest doctrines imposed on his followers. She soon became Abbess of the Benedictine Nuns with whom she was associated by the saint; and afterwards founded an order of her own—the order of " Poor Clares." The fame of her piety and humility, of her devotion to the cause of the sick, the afflicted, and the poor, spread far and wide. The most illustrious of the ecclesiastics of her time attended at her convent as at a holy shrine. Pope Innocent the Fourth visited her, as a testimony of his respect for her virtues; and paid homage to her memory when her blameless existence had closed, by making one among the mourners who followed her to the grave. Her name had been derived from the Latin word that signifies *purity;* and from first to last, her life had kept the promise of her name.

Poor St. Clare! If she could look back, with the thoughts and interests of the days of her mortality, to the world that she has quitted for ever, how sadly would she now contemplate the Holy Well which was once hallowed in her name and for her sake! But one arched wall, thickly overgrown with ivy, still remains erect in the place that the old Oratory occupied. Fragments of its roof, its cornices, and the mouldings of its windows lie scattered on the ground, half hidden by the grasses and ferns twining prettily around them. A double cross of stone stands, sloping towards the earth, at a little distance off—soon perhaps to share the fate of the prostrate ruins about it. How changed the scene here, since the time when the rural christening procession left the church, to proceed down the quiet pathway to the Holy Well—when children were baptized in the pure spring; and vows were offered up under the roof of the Oratory, and prayers were repeated before the sacred cross! These were the pious usages of a past age; these were the ceremonies of an ancient church, whose innocent and reverent custom it was to connect closer together the beauty of Nature and the beauty of Religion, by such means as the consecration of a spring, or the erection of a roadside cross. There has been something of sacrifice as well as of glory, in the effort by which we, in our time, have freed ourselves from what was superstitious and tyrannical in the faith of the times of old—it has cost us the loss of much of the better part of that faith which was not superstition, and of more which was not tyranny. The spring of St. Clare is nothing to the cottager of our day but a place to draw water from; the village lads now lounge whistling on the fallen stones, once the consecrated arches under which their humble ancestors paused on the pilgrimage, or knelt in prayer. Wherever the eye turns, all around it speaks the melancholy language of desolation and decay—all but the water of the Holy Well. Still the little pool remains the fitting type of its patron saint—pure and tranquil as in the bygone days, when the name of St. Clare was something more than the title to a village legend, and the spring of St. Clare something better than a sight for the passing tourist among the Cornish moors.*

---

* I visited St. Cleer's Well, for the second time, ten years

D

We happened to arrive at the well at the period when the villagers were going home to dinner. After the first quarter of an hour, we were left almost alone among the ruins. The only person who approached to speak to us was a poor old woman, bent and tottering with age, who lived in a little cottage hard by. She brought us a glass, thinking we might wish to taste the water of the spring; and presented me with a rose out of her garden. Such small scraps of information as she had gathered together about the well, she repeated to us in low, reverential tones, as if its former religious uses still made it an object of veneration in her eyes. After a time, she too quitted us; and we were then left quite alone by the side of the spring.

It was a bright, sunshiny day; a pure air was abroad; nothing sounded audibly but the singing of birds at some distance, and the rustling of the few leaves that clothed one or two young trees in a neighbouring garden. Unoccupied though

---

after the above lines were written; and I am happy to say that two gentlemen, interested in this beautiful ruin, are about to restore it—using the old materials for the purpose, and exactly following the original design. (March, 1861.)

I was, the minutes passed away as quickly and as unheeded with me, as with my companion who was busily engaged in sketching. The ruins of the ancient Oratory, viewed amid the pastoral repose of all things around them, began imperceptibly to exert over me that mysterious power of mingling the impressions of the present with the memories of the past, which all ruins possess. While I sat looking idly into the water of the well, and thinking of the groups that had gathered round it in years long gone by, recollections began to rise vividly on my mind of other ruins that I had seen in other countries, with friends, some scattered, some gone now—of pleasant pilgrimages, in boyish days, along the storied shores of Baiæ, or through the desolate streets of the Dead City under Vesuvius—of happy sketching excursions to the aqueducts on the plains of Rome, or to the temples and villas of Tivoli; during which, I had first learned to appreciate the beauties of Nature under guidance which, in this world, I can never resume; and had seen the lovely prospects of Italian landscape pictured by a hand now powerless in death. Remembrances such as these, of pleasures which remembrance only can recall as they were, made time fly fast for me by the brink of the holy well. I could have sat there all day, and should not have felt, at night, that the day had been ill spent.

But the sunlight began to warn us that noon was long past. We had some distance yet to walk, and many things more to see. Shortly after my friend had completed his sketch, therefore, we reluctantly left St. Clare's Well, and went on our way briskly, up the little valley, and out again on the wide surface of the moor.

It was now our object to steer a course over the wide plain around us, leading directly to the "Cheese-Wring" rocks (so called from their supposed resemblance to a Cornish cheese-press or "*wring*"). On our road to this curiosity, about a mile and a half from St. Clare's Well, we stopped to look at one of the most perfect and remarkable of the ancient British monuments in Cornwall. It is called Trevethey Stone, and consists of six large upright slabs of granite, overlaid by a seventh, which covers them in the form of a rude, slanting roof. These slabs are so irregular in form as to look quite unhewn. They all vary in size and thickness. The whole structure rises to a height, probably, of fourteen feet; and, standing as it does on elevated ground, in a barren country, with no stones of a similar kind erected near it, presents an appearance of rugged grandeur and aboriginal simplicity, which renders it an impressive, almost a startling object to look on. Antiquaries have discovered that its name signifies The Place of Graves; and have discovered no more. No inscription appears on it; the date of its erection is lost in the darkest of the dark periods of English history.

Our path had been gradually rising all the way from St. Clare's Well; and, when we left Trevethey Stone, we still continued to ascend, proceeding along the tram-way leading to the Caraton Mine. Soon the scene presented another abrupt and extraordinary change. We had been walking hitherto amid almost invariable silence and solitude; but now, with each succeeding minute, strange, mingled, unintermitting noises began to grow louder and louder around us. We followed a sharp curve in the tram-way, and immediately found ourselves saluted by an entirely new prospect, and surrounded by an utterly bewildering noise. All about us monstrous wheels were turning slowly; machinery was clanking and groaning in the hoarsest discords; invisible waters were pouring onward with a rushing sound; high above our heads, on skeleton platforms, iron chains clattered fast and fiercely over iron pulleys, and huge steam pumps puffed and gasped, and slowly raised and depressed their heavy black beams of wood. Far beneath the embankment on which we stood, men, women, and children were breaking and washing ore in a perfect marsh of copper-coloured mud and copper-coloured water. We had penetrated to the very centre of the noise, the bustle, and the population on the surface of a great mine.

When we walked forward again, we passed through a thick plantation of young firs; and then, the sounds behind us became slowly and solemnly deadened the further we went on. When we had arrived at the extremity of the line of trees, they ceased softly and suddenly. It was like a change in a dream.

We now left the tram-way, and stood again on the moor—on a wilder and lonelier part of it than we had yet beheld. The Cheese-Wring and its adjacent rocks were visible a mile and a half away, on the summit of a steep hill. Wherever we looked, the horizon was bounded by the long, dark, undulating edges of the moor. The ground rose and fell in little hillocks and hollows, tufted with dry grass and furze, and strewn throughout with fragments of granite.

The whole plain appeared like the site of an ancient city of palaces, overthrown and crumbled into atoms by an earthquake. Here and there, some cows were feeding; and sometimes a large crow winged his way lazily before us, lessening and lessening slowly in the open distance, until he was lost to sight. No human beings were discernible anywhere; the majestic loneliness and stillness of the scene were almost oppressive both to eye and ear. Above us, immense fleecy masses of brilliant white cloud, wind-driven from the Atlantic, soared up grandly, higher and higher over the bright blue sky. Everywhere, the view had an impressively stern, simple, aboriginal look. Here were tracts of solitary country which had sturdily retained their ancient character through centuries of revolution and change; plains pathless and desolate even now, as when Druid processions passed over them by night to the place of the secret sacrifice, and skin-clad warriors of old Britain halted on them in council, or hurried across them to the fight.

On we went, up and down, in a very zigzag course, now looking forward towards the Cheese-Wring from the top of a rock, now losing sight of it altogether in the depths of a hollow. By the time we had advanced about half way over the distance it was necessary for us to walk, we observed, towards the left hand, a wide circle of detached upright rocks. These we knew, from descriptions and engravings, to be the "Hurlers"— so we turned aside at once to look at them from a nearer point of view.

There are two very different histories of these rocks; the antiquarian account of them is straightforward and practical enough, simply asserting that they are the remains of a Druid temple, the whole region about them having been one of the principal stations of the Druids in Cornwall. The popular account of the Hurlers (from which their name is derived) is very different. It is contended, on the part of the people, that once upon a time (nobody knows how long ago), these rocks were Cornish men, who profanely went out (nobody knows from what place), to enjoy the national sport of hurling the ball on one fine "Sabbath morning," and were suddenly turned into pillars of stone, as a judgment on their own wickedness, and a warning to all their companions as well.

Having to choose between the antiquarian hypothesis and the popular legend on the very spot to which both referred, a common susceptibility to the charms of romance at once determined us to pin our faith on the legend. Looking at the Hurlers, therefore, in the peculiar spirit of the story attached to them, as really and truly petrified ball-players, we observed, with great interest, that some of them must have been a little above, and others a little below our own height, in their lifetime; that some must have been very corpulent, and others very thin persons; that one of them, having a protuberance on his head remarkably like a night-cap in stone, was possibly a sluggard as well as a Sabbath-breaker, and might have got out of his bed just in time to "hurl;" that another, with some faint resemblance left of a fat grinning human face, leaned considerably out of the perpendicular, and was, in all probability, a hurler of intemperate habits. At some distance off we remarked a high stone standing entirely by itself, which, in the absence of any positive information on the subject, we presumed to consider as the petrified effigy of a tall man who ran after the ball. In the opposite direction, other stones were dotted about irregularly, which we could only imagine to represent certain misguided wretches who had attended as spectators of the sports, and had therefore incurred the same penalty as the hurlers themselves. These humble results of observations taken on the spot, may possibly be useful, as tending to offer some startling facts from ancient history to the next pious layman in the legislature who gets up to propose the next series of Sabbath prohibitions for the benefit of the profane laymen in the nation.

Abandoning any more minute observation of the Hurlers than that already recorded, in order to husband the little time still left to us, we soon shaped our course again in the direction of the Cheese-Wring. We arrived at the base of the hill on which it stands, in a short time and without any difficulty; and beheld above us a perfect chaos of rocks piled up the entire surface of the eminence. All the granite we had seen before was as nothing compared with the granite we now looked on. The masses were at one place heaped up in great irregular cairns—at another, scattered confusedly over the ground; poured all along in close, craggy lumps; flung about hither and thither, as if in reckless sport, by the hands of giants. Above the whole, rose the weird fantastic form of the Cheese-Wring, the wildest and most wondrous of all the wild and wondrous structures in the rock architecture of the scene.

If a man dreamt of a great pile of stones in a nightmare, he would dream of such a pile as the

Cheese-Wring. All the heaviest and largest of the seven thick slabs of which it is composed are at the top; all the lightest and smallest at the bottom. It rises perpendicularly to a height of thirty-two feet, without lateral support of any kind. The fifth and sixth rocks are of immense size and thickness, and overhang fearfully, all round, the four lower rocks which support them, All are perfectly irregular; the projections of one do not fit into the interstices of another; they are heaped up loosely in their extraordinary top-heavy form, on slanting ground half-way down a steep hill. Look at them from whatever point you choose, there is still all that is heaviest, largest, strongest, at the summit, and all that is lightest, smallest, weakest, at the base. When you first see the Cheese-Wring, you instinctively shrink from walking under it. Beholding the tons on tons of stone balanced to a hair's breadth on the mere fragments beneath, you think that with a pole in your hand, with one push against the top rocks, you could hurl down the hill in an instant a pile which has stood for centuries, unshaken by the fiercest hurricane that ever blew, rushing from the great void of an ocean over the naked surface of a moor.

Of course, theories advanced by learned men are

not wanting to explain such a phenomenon as the Cheese-Wring. Certain antiquaries have undertaken to solve this curious problem of Nature in a very off-hand manner, by asserting that the rocks were heaped up as they now appear, by the Druids, with the intention of astonishing their contemporaries and all posterity by a striking exhibition of their architectural skill. (If any of these antiquarian gentlemen be still living, I would not recommend them to attempt a practical illustration of their theory by building miniature Cheese-Wrings out of the contents of their coal-scuttles!) The second explanation of the extraordinary position of the rocks is a geological explanation, and is apparently the true one. It is assumed on this latter hypothesis, that the Cheese-Wring, and all the adjacent masses of stone, were once covered, or nearly covered, by earth, and were thus supported in an upright form; that the wear and tear of storms gradually washed away all this earth, from between the rocks, down the hill, and then left such heaps of stones as were accidentally complete in their balance on each other, to stand erect, and such as were not, to fall flat on the surface of the hill in all the various positions in which they now appear. Accepting this theory as the right one, it still seems strange that there should be only one Cheese-Wring on the hill—but so it is. Plenty of rocks are to be seen there piled one on another; but none of them are piled in the same extraordinary manner as the Cheese-Wring, which stands alone in its grandeur, a curiosity that even science may wonder at, a sight which is worth a visit to Cornwall, if Cornwall presented nothing else to see.

Besides the astonishment which the rock scenery on the hill was calculated to excite, we found in its neighbourhood an additional cause for surprise of a very different description. Just as we were preparing to ascend the eminence, the silence of the great waste around us was broken by a long and hearty cheer. The Hurlers themselves, if they had suddenly returned to a state of flesh and blood, and resumed their interrupted game, could hardly have made more noise, or exhibited a greater joviality of disposition, than did some three or four tradesmen of the town of Liskeard, who had been enjoying a pic-nic under the Cheese-Wring, had seen us approaching over the plain, and now darted out of their ambush to welcome us, flourishing porter-bottles in their hands as olive branches of peace, amity, and good-will. My companion skilfully contrived to make his escape ; but

I was stopped and surrounded in an instant. One benevolent stranger held a glass in a very slanting position, while a brother philanthropist violently uncorked a bottle and directed half of its contents in a magnificent jet of light brown froth all over everybody, before he found the way into the tumbler. It was of no use to decline imbibing the remainder of the light brown froth—"*There* was the Cheese-Wring (cried all the benevolent strangers in chorus), and *here* was the porter—*I* must drink all their good healths, and *they* would all drink mine—this was Cornish hospitality, and Cornish hospitality was notoriously the finest thing in the world! As for my friend there, who was drawing, they bore him no ill-will because he wouldn't drink—they would buy his drawing, and one of the commercial gentlemen, who was a stationer, would publish a hundred, two hundred, five hundred, a thousand copies of it, on sheets of letter-paper, price one penny! What had I got to say to that?—If that wasn't hospitality, what the devil was?"

All this might have been very amusing, and our new friends might have proved excellent companions, under a different set of circumstances. But, as things were, we neither of us felt at all sorry when their manners subsequently exhibited a slight change, under the influence of further potations of porter. Soon, they began to look stolid and suspicious—suddenly, they discovered that we were not quite such good company as they had thought us at first—finally, they took their departure in solemn silence, leaving us free at last to mount the hill, and look out uninterruptedly on the glorious view from the summit, which extended over a circumference of a hundred miles.

Turning our faces towards the north-east, and standing now on the topmost rock of one of the most elevated situations in Cornwall, we were able to discern the sea on either side of us. Two faint lines of the softest, haziest blue, indicated the Bristol Channel on the one hand, and the English Channel on the other. Before us lay a wide region of downs and fields, all mapped out in every variety of form by their different divisions of wall and hedge-row—while, farther away yet, darker and more indefinite, appeared the Dartmoor forest and the Dartmoor hills. It was just that hour before the evening, at which the atmosphere acquires a more mellow purity, a more perfect serenity and warmth, than at earlier periods of the day. The shadows of great clouds lay in vast lovely shapes of purple blue over the whole visible tract of country, contrasting in exquisite beauty with the sunny glimpses of landscape shining between them. Beneath us, the picturesque confusion of rocks, topped by the quaint form of the Cheese-Wring, seemed to fade away mysteriously into the grass of the moorland; beyond which, high up where the hills rose again, a little lake, called Dosmery Pool, shone in the sunlight with dazzling, diamond brightness. In the opposite direction, towards the west, the immediate prospect was formed by the rugged granite ridges, towering one behind the other, of Sharp Torr and Kilmarth—the long hazy outlines of the plains and hill-tops of southern and inland Cornwall closing grandly the distant view.

All that we had hitherto seen on and around the spot where we now stood, had not yet exhausted its objects of attraction for strangers. Descending the rocks in a new direction, after taking a last look at the noble prospect visible from their summit, we proceeded to a particular spot near the base of the hill, where the granite was scattered in remarkable abundance. Our purpose here was to examine some stones which are well known to all the quarrymen in the district, as associated with an extraordinary story and an extraordinary man.

During the earlier half of the last century, there lived in one of the villages on the outskirts of the moor on which the Cheese-Wring stands, a stone-cutter named Daniel Gumb. This man was noted among his companions for his taciturn eccentric character, and for his attachment to mathematical studies. Such leisure time as he had at his command he devoted to pondering over the problems of Euclid: he was always drawing mysterious complications of angles, triangles, and parallelograms, on pieces of slate, and on the blank leaves of such few books as he possessed. But he made very slow progress in his studies. Poverty and hard work increased with the increase of his family, and obliged him to give up his mathematics altogether. He laboured early and laboured late; he hacked and hewed at the hard material out of which he was doomed to cut a livelihood, with unremitting diligence; but times went so ill with him, that in despair of ever finding them better, he took a sudden resolution of altering his manner of living, and retreating from the difficulties that he could not overcome. He went to the hill on which the Cheese-

E

Wring stands, and looked about among the rocks until he found some that had accidentally formed themselves into a sort of rude cavern. He widened this recess; he propped up a great wide slab, to make its roof: he cut out in a rock that rose above this, what he called his bed-room—a mere longitudinal slit in the stone, the length and breadth of his body, into which he could roll himself sideways when he wanted to enter it. After he had completed this last piece of work, he scratched the date of the year of his extraordinary labours (1735) on the rock; and then removed his wife and family from their cottage, and lodged them in the cavity he had made—never to return during his lifetime to the dwellings of men!

Here he lived and here he worked, when he could get work. He paid no rent now: he wanted no furniture; he struggled no longer to appear to the world as his equals appeared; he required no more money than would procure for his family and himself the barest necessaries of life; he suffered no interruptions from his fellow-workmen, who thought him a madman, and kept out of his way; and—most precious privilege of his new position—he could at last shorten his hours of labour, and lengthen his hours of study, with impunity. Having no temptations to spend money, no hard demands of an inexorable landlord to answer, he could now work with his brains as well as his hands; he could toil at his problems, scratching them upon the tops of rocks, under the open sky, amid the silence of the great moor. Henceforth, nothing moved, nothing depressed him. The storms of winter rushed over his unsheltered dwelling, but failed to dislodge him. He taught his family to brave solitude and cold in the cavern among the rocks, as *he* braved them. In the cell that he had scooped out for his wife (the roof of which has now fallen in) some of his children died, and others were born. They point out the rock where he used to sit on calm summer evenings, absorbed over his tattered copy of Euclid. A geometrical "puzzle," traced by his hand, still appears on the stone. When he died, what became of his family, no one can tell. Nothing more is known of him than that he never quitted the wild place of his exile; that he continued to the day of his death to live contentedly with his wife and children, amid a civilized nation, under such a shelter as would hardly serve the first savage tribes of the most savage country—to live, starving out poverty and want on a barren wild; forsaking all things enduring all things for the love of Knowledge, which he could still nobly follow through trials and extremities, without encouragement of fame or profit, without vantage ground of station or wealth, for its own dear sake. Beyond this, nothing but conjecture is left. The cell, the bed-place, the lines traced on the rocks, the inscription of the year in which he hewed his habitation out of them, are all the memorials that remain of Daniel Gumb.

We lingered about the wild habitation of the stonemason and his family, until sunset. Long shadows of rocks lay over the moor, the breeze had freshened and was already growing chill, when we set forth, at last, to trace our way back to Liskeard. It was too late now to think of proceeding on our journey, and sleeping at the next town on our line of route.

Returning in a new direction, we found ourselves once more walking on a high road, just as the sun had gone down, and the grey twilight was falling softly over the landscape. Stopping near a lonely farm-house, we went into a field to look at another old British monument to which our attention had been directed. We saw a square stone column—now broken into two pieces—ornamented with a curiously carved pattern, and exhibiting an inscription cut in irregular, mysterious characters. Those who have deciphered them, have discovered that the column is nearly a thousand years old; that it was raised as a sepulchral monument over the body of Dungerth King of Cornwall; and that the letters carved on it form some Latin words, which may be thus translated:—"PRAY FOR THE SOUL OF DUNGERTH." Seen in the dim light of the last quiet hour of evening, there was something solemn and impressive about the appearance of the old tombstone—simple though it was. After leaving it, we soon entered once more into regions of fertility. Cottages, corn-fields, and trees surrounded us again. We passed through pleasant little valleys; over brooks crossed by quaint wooden bridges; up and down long lanes, where tall hedges and clustering trees darkened the way—where the stag-beetle flew slowly by, winding "his small but sullen horn," and glow-worms glimmered brightly in the long, dewy grass by the roadside. The moon, rising at first red and dull in a misty sky, brightened as we went on, and lighted us brilliantly along all that remained of our night-walk back to the town.

I have only to add, that, when we arrived at

Liskeard, the lachrymose landlady of the inn benevolently offered us for supper the identical piece of cold "*corned beef*" which she had offered us for dinner the day before; and further proposed that we should feast at our ease in the private dungeon dining-room at the back of the house. But one mode of escape was left—we decamped at once to the large and comfortable hotel of the town; and there our pleasant day's pilgrimage to the moors of Cornwall concluded as agreeably as it had begun.

IV.

CORNISH PEOPLE.

It is my purpose, in this place, to communicate some few facts relating to the social condition of the inhabitants of Cornwall, which were kindly furnished to me by friends on the spot; adding to the statement thus obtained, such anecdotes and illustrations of popular character as I collected from my own observations in the capacity of a tourist on foot.

If the reader desires to compare at a glance the condition of the Cornish people with the condition of their brethren in other parts of England, one small particle of practical information will enable him to do so at once. In the Government Tables of Mortality for Cornwall there are no returns of death from starvation.

Many causes combine to secure the poor of Cornwall from that last worst consequence of poverty to which the poor in most of the other divisions of England are more or less exposed. The number of inhabitants in the county is stated by the last census at 341,269—the number of square miles that they have to live on, being 1327.*—This will be found on proper computation and comparison, to be considerably under the average population of a square mile throughout the rest of England. Thus, the supply of men for all purposes does not appear to be greater than the demand in Cornwall. The remote situation of the county guarantees it against any considerable influx of strangers to compete with the natives for work on their own ground. We met a farmer there, who was so far from being besieged in harvest time by claimants for labour on his land, that he was obliged to go forth to seek them himself at a neighbouring town, and was doubtful whether he should find men enough left him unemployed at the mines and the fisheries, to gather in his crops

---

* It may be necessary to remind the reader that this statement respecting the population of Cornwall was written in the year 1850. I have no means at my disposal of ascertaining what the increase in numbers may have been during the last ten years.—(March, 1861.)

in good time at two shillings a day and as much "victuals and drink" as they cared to have.

Another cause which has contributed, in some measure, to keep Cornwall free from the burthen of a surplus population of working men must not be overlooked. Emigration has been more largely resorted to in that county, than perhaps in any other in England. Out of the population of the Penzance Union alone, nearly five per cent. left their native land for Australia, or New Zealand, in 1849. The potato-blight was, at that time, assigned as the chief cause of the readiness to emigrate; for it damaged seriously the growth of a vegetable, from the sale of which, at the London markets, the Cornish agriculturalists derived large profits, and on which (with their fish) the Cornish poor depend as a staple article of food.

It is by the mines and fisheries (of both of which I shall speak particularly in another place) that Cornwall is compensated for a soil, too barren in many parts of the county, to be ever well cultivated except at such an expenditure of capital as no mere farmer can afford. From the inexhaustible mineral treasures in the earth, and from the equally inexhaustible shoals of pilchards which annually visit the coast, the working population of Cornwall derive their regular means of support, where agriculture would fail them. At the mines, the regular rate of wages is from forty to fifty shillings a month; but miners have opportunities of making more than this. By what is termed "working on tribute," that is, agreeing to excavate the mineral lodes for a per centage on the value of the metal they raise, some of them have been known to make as much as six and even ten pounds each, in a month. When they are unlucky in their working speculations, or perhaps thrown out of employment altogether by the shutting up of a mine, they still have a fair opportunity of obtaining farm labour, which is paid for (out of harvest time) at the rate of nine shillings a week. But this is a resource of which they are rarely obliged to take advantage. A plot of common ground is included with the cottages that are let to them; and the cultivation of this, helps to keep them and their families, in bad times, until they find an opportunity of resuming work; when they may perhaps make as much in one month, as an agricultural labourer can in twelve.

The fisheries not only employ all the inhabitants of the coast, but, in the pilchard season, many of the farm work-people as well. Ten thousand persons—men, women, and children — derive their regular support from the fisheries; which are so amazingly productive, that the "drift," or deep-sea fishing, in Mount's Bay alone, is calculated to realize, on the average, 30,000*l.* per annum.

To the employment thus secured for the poor in the mines and fisheries is to be added, as an advantage, the cheapness of rent and living in Cornwall. Good cottages are let at from fifty shillings, to between three and four pounds a-year — turf for firing grows in plenty on the vast tracts of common land overspreading the country—all sorts of vegetables are abundant and cheap, with the exception of potatoes, which so decreased in 1849, in consequence of the disease, that the winter stock was imported from France, Belgium, and Holland. The early potatoes, however, grown in May and June, are cultivated in large quantities, and realize on exportation a very high price. Corn generally sells a little above the average. Fish is always within the reach of the poorest people. In a good season, a dozen pilchards are sold for one penny. Happily for themselves, the poor in Cornwall do not partake the senseless prejudice against fish, so obstinately adhered to by the poor in many other parts of England. A Cornishman's national pride is in his pilchards— he likes to talk of them, and boast about them to strangers; and with reason, for he depends for the main support of life on the tribute of these little fish which the sea yields annually in almost countless shoals.

The workhouse system in Cornwall is said, by those who are well qualified to form an opinion on the subject, to be generally well administered; the Unions in the eastern part of the county being the least stringent in their regulations, and the most liberal in giving out-of-door relief.

Such, briefly, but I think not incorrectly stated, is the condition of the poor in Cornwall, in relation to their means of subsistence as a class. Looking to the fact that the number of labourers there is not too much for the labour; comparing the rate of wages with rent and the price of provisions; setting the natural advantages of the county fairly against its natural disadvantages, it is impossible not to conclude that the Cornish poor suffer less by their poverty, and enjoy more opportunities of improving their social position, than the majority of their brethren in many other counties of England. The general demeanour and language of the people themselves amply warrant this conclusion. The Cornish are essentially a cheerful, contented race. The views of the working men are remarkably moderate and sensible—I never met with so few grumblers anywhere.

My opportunities of correctly estimating the state of education among the people, were not sufficiently numerous to justify me in offering to the reader more than a mere opinion on the subject. Such few observations as I was able to make, inclined me to think that, in education, the mass of the population was certainly below the average in England, with one exception—that of the classes employed in the mines. All of these men with whom I held any communication, would not have been considered badly-informed persons in a higher condition of life. They possessed much more than a common mechanical knowledge of their own calling, and even showed a very fair share of information on the subject of the history and antiquities of their native county. As usual, the agricultural inhabitants appeared to rank lowest in the scale of education and general intelligence. Among this class, and among the fishermen, the strong superstitious feelings of the ancient days of Cornwall still survive, and promise long to remain, handed down from father to son as heirlooms of tradition, gathered together in a remote period, and venerable in virtue of their antiquity. The notion, for instance, that no wound will fester as long as the instrument by which it was inflicted is kept bright and clean, still prevails extensively among them. But a short time since, a boy in Cornwall was placed under the care of a medical man (who related the anecdote to me) for a wound in the back from a pitchfork; his relatives—cottagers of respectability—firmly believe that his cure was accelerated by the pains they took to keep the prongs of the pitchfork in a state of the highest polish, night and day, throughout the whole period of his illness, and down to the last hour of his complete restoration to health.

Another and a more remarkable instance of the superstitions prevailing among the least educated classes of the people, was communicated to me by the same informant—a gentleman whose life had been passed in Cornwall, and who was highly and deservedly respected by all those among whom he resided.*

---

* The gentleman here referred to—whose kind assistance while I was writing these pages I can never forget—was Mr. Richard Moyle, long resident as a medical man at Penzance.

A small farmer living in one of the most western districts of the county, died some years back of what was supposed at the time to be "English Cholera." A few weeks after his decease, his wife married again. This circumstance excited some attention in the neighbourhood. It was remembered that the woman had lived on very bad terms with her late husband, that she had on many occasions exhibited strong symptoms of possessing a very vindictive temper, and that during the farmer's life-time she had openly manifested rather more than a Platonic preference for the man whom she subsequently married. Suspicion was generally excited: people began to doubt whether the first husband had died fairly. At length the proper order was applied for, and his body was disinterred. On examination, enough arsenic to have poisoned three men was found in his stomach. The wife was accused of murdering him, was tried, convicted on the clearest evidence, and hanged. Very shortly after she had suffered capital punishment, horrible stories of a ghost were widely circulated. Certain people declared that they had seen a ghastly resemblance of

---

Since my first visit to Cornwall, death has removed Mr. Moyle from the scene of his labours, to the lasting and sincere regret of all who knew him.—(March, 1861.)

the murderess, robed in her winding-sheet, with the black mark of the rope round her swollen neck, standing on stormy nights upon her husband's grave, and digging there with a spade in hideous imitation of the actions of the men who had disinterred the corpse for medical examination. This was fearful enough—nobody dared go near the place after night-fall. But soon, another circumstance was talked of, in connexion with the poisoner, which affected the tranquillity of people's minds in the village where she had lived, and where it was believed she had been born, more seriously than even the ghost-story itself.

Near the church of this village there was a well, celebrated among the peasantry of the district for one remarkable property—every child baptized in its water (with which the church was duly supplied on christening occasions) was secure from ever being hanged. No one doubted that all the babies fortunate enough to be born and baptized in the parish, though they might live to the age of Methuselah, and might during that period commit all the capital crimes recorded in the "Newgate Calendar," were still destined to keep quite clear of the summary jurisdiction of Jack Ketch—no one doubted this, until the story of the apparition of the murderess began to be spread abroad. Then, awful misgivings arose in the popular mind. A woman who had been born close by the magical well, and who had therefore in all probability been baptized in its water like her neighbours of the parish, had nevertheless been publicly and unquestionably hanged. However, probability was not always truth—everybody determined that the baptismal register of the poisoner should be sought for, and that it should be thus officially ascertained whether she had been christened with the well water, or not. After much trouble, the important document was discovered—not where it was first looked after, but in a neighbouring parish vestry. A mistake had been made about the woman's birthplace—she had not been baptized in the local church, and had therefore not been protected by the marvellous virtue of the local water. Unutterable was the joy and triumph of this discovery throughout the village—the wonderful character of the parish well was wonderfully vindicated—its celebrity immediately spread wider than ever. The peasantry of the neighbouring districts began to send for the renowned water before christenings; and many of them actually continue, to this day, to bring it corked up in bottles to their churches, and to beg particularly

F

that it may be used whenever they present their children to be baptized.

Such instances of superstition as this—and others equally true might be quoted—afford, perhaps, of themselves, the best evidence of the low state of education among the people from whom they are produced. It is, however, only fair to state, that children in Cornwall are now enabled to partake of advantages which were probably not offered to their parents. Good National Schools are in operation everywhere, and are—as far as my own inquiries authorize me to report—well attended by pupils recruited from the ranks of the poorest classes.

Of the social qualities of the Cornish all that can be written may be written conscientiously in terms of the highest praise. Travelling as my companion and I did—in a manner which (whatever it may be now) was, ten years since, perfectly new to the majority of the people—we found constant opportunities of studying the popular character in its every day aspects. We perplexed some, we amused others: here, we were welcomed familiarly by the people, as travelling pedlars with our packs on our backs; there, we were curiously regarded at an awful distance, and respectfully questioned in circumlocutory phrases as to our secret designs in walking through the country. Thus, viewing us sometimes as their equals, sometimes as mysteriously superior to them, the peasantry unconsciously exhibited many of their most characteristic peculiarities without reserve. We looked at the spectacle of their social life from the most searching point of view, for we looked at it from behind the scenes.

The manners of the Cornish of all ranks, down to the lowest, are remarkably distinguished by courtesy —a courtesy of that kind which is quite independent of artificial breeding, and which proceeds solely from natural motives of kindness and from an innate anxiety to please. Few of the people pass you without a salutation. Civil questions are always answered civilly. No propensity to jeer at strangers is exhibited—on the contrary, great solicitude is displayed to afford them any assistance that they may require; and displayed, moreover, without the slightest appearance of a mercenary motive. Thus, if you stop to ask your way, you are not merely directed for a mile or two on, and then told to ask again; but directed straight to the end of your destination, no matter how far off. Turnings to the right, and turnings to the left, short cuts across moors five miles away, churches that you must keep on this hand, and rocks that you must keep on that, are impressed upon your memory with the most laborious minuteness, and shouted after you over and over again as long as you are within hearing. If the utmost anxiety to give the utmost quantity of good advice could always avail against accident or forgetfulness, no traveller in Cornwall who asks his way as he goes, need ever lose himself.

When people possess the virtue of natural courtesy they are seldom found wanting in other higher virtues that are akin to it. Household affection, ready hospitality, and great gratitude for small rewards of services rendered, are all to be found among the Cornish peasantry. Their fondness for their children is very pleasant to see. A word of inquiry or praise addressed to the mother makes her face glow with delight, and sends her away at once in search of the missing members of her little family, who are ranged before you triumphantly, with smoothed hair and carefully wiped faces, ready to be reviewed in a row. Both father and mother often wish you, at parting, a good wife and a large family (if you are not married already), just as they wish you a pleasant journey and a prosperous return home again.

Of Cornish hospitality we experienced many proofs, one of which may be related as a sample. Arriving late at a village, in the far west of the county, we found some difficulty in arousing the people of the inn. While we were waiting at the door, we heard a man who lived in a cottage near at hand, and of whom we had asked our way on the road, inquiring of some female member of his family, whether she could make up a spare bed. We had met this man proceeding in our direction, and had so far outstripped him in walking, that we had been waiting outside the inn about a quarter of an hour before he got home. When the woman answered his question in the negative, he directed her to put clean sheets on his own bed, and then came out to tell us that if we failed to obtain admission at the public-house, a lodging for the night was ready for us under his own roof. We found on inquiry, afterwards, that he had looked out of window, after getting home, while we were still disturbing the village by a continuous series of assaults on the inn door; had recognised us in the moonlight; and had thereupon not only offered us his bed, but had got out of it himself to do so. When we finally succeeded in gaining admittance to the inn, he declined an invitation to sup with us, and wishing us a good night's rest, returned to his home. I should mention, at the same time, that another bed was offered to us at the vicarage, by the clergyman of the parish; and that after this gentleman had himself seen that we were properly accommodated by our landlady, he left us with an invitation to breakfast with him the next morning. Thus is hospitality practised in Cornwall—a county where, it must be remembered, a stranger is doubly a stranger, in relation to provincial sympathies; where the national feeling is almost entirely merged in the local feeling; where a man speaks of himself as *Cornish* in much the same spirit as a Welshman speaks of himself as Welsh.

In like manner, another instance drawn from my own experience, will best display the anxiety which we found generally testified by the Cornish poor to make the best and most grateful return in their power for anything which they considered as a favour kindly bestowed. Such little anecdotes as I here relate in illustration of popular character, cannot, I think, be considered trifling; for it is by trifles, after all, that we gain our truest appreciation of the marking signs of good or evil in the dispositions of our fellow-beings; just as in the beating of a single artery under the touch, we discover an indication of the strength or weakness of the whole vital frame.

On the granite cliffs at the Land's End I met

with an old man, seventy-two years of age, of whom I asked some questions relative to the extraordinary rocks scattered about this part of the coast. He immediately opened his whole budget of local anecdotes, telling them in a quavering high-treble voice, which was barely audible above the dash of the breakers beneath, and the fierce whistling of the wind among the rocks around us. However, the old fellow went on talking incessantly, hobbling along before me, up and down steep paths and along the very brink of a fearful precipice, with as much coolness as if his sight was as clear and his step as firm as in his youth. When he had shown me all that he could show, and had thoroughly exhausted himself with talking, I gave him a shilling at parting. He appeared to be perfectly astonished by a remuneration which the reader will doubtless consider the reverse of excessive; thanked me at the top of his voice; and then led me, in a great hurry, and with many mysterious nods and gestures, to a hollow in the grass, where he had spread on a clean pocket-handkerchief a little stock-in-trade of his own, consisting of barnacles, bits of rock and ore, and specimens of dried seaweed. Pointing to these, he told me to take anything I liked, as a present in return for what I had given him. He would not hear of my buying anything; he was not, he said, a regular guide, and I had paid him more already than such an old man was worth—what I took out of his handkerchief I must take as a present only. I saw by his manner that he would be really mortified if I contested the matter with him, so as a present I received one of his pieces of rock—I had no right to deny him the pleasure of doing a kind action, because there happened to be a few more shillings in my pocket than in his.

Nothing can be much better adapted to show how simple and unsophisticated the Cornish character still remains in many respects, than Cornish notions of organizing a public festival, and Cornish enjoyment of that festival when it is organized. We had already seen how they managed a public boat-race at Looe, and we saw again how they conducted the preparations for the same popular festival, on a larger scale, at the coast town of Fowey.

In the first place, the dormant public enthusiasm was stimulated by music at an uncomfortably early hour in the morning. Two horn players and a clarionet player; a fat musician who blew through a very small fife and kept time with his head; and a withered little man who beat furiously on a mighty drum—drew up in martial array, one behind the other, before the principal inn. Two boys, staring about them in a stolidly important manner, and carrying flags which bore a suspicious resemblance to India pocket handkerchiefs sewn together, formed in front of the musicians. Two corpulent, solemn, elderly gentlemen in black (belonging, apparently, to the churchwarden-type of the human species), formed in their turn on each side of the boys—and then the procession started; walking briskly up and down, and in and out, and round and round the same streets, over and over again; the musicians playing on all their instruments at once (drum included), without a moment's intermission on the part of any one of them. Nothing could exceed the gravity and silence of the popular concourse which followed this grotesque procession. The solemn composure on the countenances of the two corpulent civil officers who went before it, was reflected on the features of the smallest boy who followed humbly behind. Profound musical amateurs in attendance at a classical quartett concert, could have exhibited no graver or more breathless attention than that displayed by the inhabitants of Fowey, as they marched at the heels of the peripatetic town-band.

But, while the music was proceeding, another adjunct to the dignity of the festival was in course of preparation, which appealed more strongly to popular sympathy even than the band and procession. A quantity of young trees—miserable little saplings cut short in their early infancy—were brought into the town, curiously sharpened at the stems. Holes were rapidly drilled in the ground, here, there, and everywhere, for their reception, at corners of house walls. While men outside set them up, women in a high state of excitement appeared at first-floor windows with long pieces of string, which they fastened to the branches to steady the trees at the top, hauling them about this way and that most unmercifully during the operation, and then vanishing to tie the loose ends of the lines to bars of grates and legs of tables. Mazes of long tight strings ran all across our room at the inn; broken twigs and drooping leaves peered in sadly at us through the three windows that lighted it. We were driven about from corner to corner out of the way of this rigging by an imperious old woman, who fastened and fettered the wretched trees with as fierce an air as if they were criminals whom she was handcuffing, and who at last fairly told us that she thought we had better leave the room, and see how beautiful things looked from the outside. On obeying this intimation, we found that the trees had absorbed the whole public attention to themselves. The band marched by, playing furiously; but the boys deserted it. The people from the country, hastening into the town, hot and eager, paused, reckless of the music, reckless of the flags, reckless of the procession, to look forth upon the streets "with verdure clad." The popularity of the Sons of Apollo was a thing of the past already! Nothing can well be imagined more miserably ugly than the appearance of the trees, standing strung into unnatural positions, and looking half dead already; but they evidently inspired the liveliest public satisfaction. Women returned to the windows to give a last perfecting tug to their branches; men patted approvingly with spades the loose earth round their stems. Spectators, one by one, took a near view and a distant view, and then walked gently by and took an occasional view, and lastly gathered together in little groups and took a general view. As connoisseurs look at their pictures, as mothers look at their children, as lovers look at their mistresses—so did the people of Fowey assemble with one accord and look at their trees.

After all, however, I shall perhaps best illustrate

the simplicity of character displayed by the Cornish country-people, if I leave the less amusing preparations for inaugurating the Fowey boat-race untold, and describe some of the peculiarities of behaviour and remark which the appearance of my companion and myself called forth in all parts of Cornwall. The mere sight of two strangers walking with such appendages as knapsacks strapped on their shoulders, seemed of itself to provoke the most unbounded wonder. We were stared at with almost incredible pertinacity and good humour. People hard at work, left off to look at us; while groups congregated at cottage doors, walked into the middle of the road when they saw us approach, looked at us in front from that commanding point of view until we passed them, and then wheeled round with one accord and gazed at us behind as long as we were within sight. Little children ran in-doors to bring out large children, as we drew near. Farmers, overtaking us on horseback, pulled in, and passed at a walk, to examine us at their ease. With the exception of bedridden people and people in prison, I believe that the whole population of Cornwall looked at us all over—back view and front view—from head to foot!

This staring was nowhere accompanied, either on

the part of young or old, by a jeering word or an impertinent look. We evidently astonished the people, but we never tempted them to forget their natural good-nature, forbearance, and self-restraint. On our side, the attentive scrutiny to which we were subjected, was at first not a little perplexing. It was difficult not to doubt occasionally whether some unpleasantly remarkable change had not suddenly taken place in our personal appearance—whether we might not have turned green or blue on our travels, or have got noses as long as the preposterous nose of the traveller through Strasburgh, in the tale of Slawkenbergius. It was not until we had been some days in the county that we began to discover, by some such indications as the following, that we owed the public attention to our knapsacks, and not to ourselves.

We enter a small public-house by the roadside to get a draught of beer. In the kitchen, we behold the landlord and a tall man who is a customer. Both stare as a matter of course; the tall man especially, after taking one look at our knapsacks, fixes his eyes firmly on us and sits bolt upright on the bench without saying a word—he is evidently prepared for the worst we can do. We get into conversation with the landlord, a jovial, talkative fellow, who desires greatly to know what we are, if we have no objection. We ask him, what he thinks we are?—" Well," says the landlord, pointing to my friend's knapsack, which has a square ruler strapped to it, for architectural drawing—" well, I think you are both of you *mappers* —mappers who come here to make new roads—you may be coming to make a railroad, I dare say— we've had mappers in the country before this—I know a mapper myself—here's both your good healths!" We drink the landlord's good health in return, and disclaim the honour of being " mappers ;" we walk through the country (we tell him) for pleasure alone, and take any roads we can get, without wanting to make new ones. The landlord would like to know, if that is the case, why we carry those weights at our backs?—Because we want to take our luggage about with us. Couldn't we pay to ride?—Yes, we could. And yet we like walking better?—Yes we do. This last answer utterly confounds the tall customer, who has been hitherto listening intently to the dialogue. It is evidently too much for his credulity—he pays his reckoning, and walks out in a hurry without uttering a word. The landlord appears to be convinced, but it is only in appearance. We leave him standing at his door, keeping his eye on us as long as we are in sight, still evidently persuaded that we are "mappers," but "mappers" of a bad order whose presence is fraught with some unknown peril to the security of the Queen's highway.

We get on into another district. Here, public opinion is not flattering. Some of the groups, gathered together in the road to observe us, begin to speculate on our characters before we are quite out of hearing. Then, this sort of dialogue, spoken in serious, subdued tones, just reaches us: Question—What can they be? Answer—"*Trodgers!*"

This is particularly humiliating, because it happens to be true. We certainly do trudge, and are therefore properly, though rather unceremoniously, called trudgers, or "trodgers." But we sink to a lower depth yet, a little further on. We are viewed as objects for pity. It is a fine evening; we stop and lean against a bank by the roadside to look at the sunset. An old woman comes tottering by on high pattens, very comfortably and nicely clad. She sees our knapsacks, and instantly stops in front of us, and begins to moan lamentably. Not understanding at first what this means, we ask respectfully if she feels at all ill? "Ah, poor fellows! poor fellows!" she sighs in answer, "obliged to carry all your baggage on your own backs!—very hard! poor lads! very hard, indeed!" And the good old soul goes away groaning over our evil plight, and mumbling something which sounds very like an assurance that she has got no money to give us.

In another part of the county we rise again gloriously in worldly consideration. We pass a cottage; a woman looks out after us, over the low garden wall, and rather hesitatingly calls us back. I approach her first, and am thus saluted: "If you please, sir, what have you got to sell?" Again, an old man meets us on the road, stops, cheerfully taps our knapsacks with his stick, and says: "Aha! you're tradesmen, eh? things to sell? I say, have you got any tea" (pronounced *tay*); "I'll buy some *tay!*" Further on, we approach a group of miners breaking ore. As we pass by, we hear one asking amazedly, "What have they got to sell in those things on their backs?" and another answering, in the prompt tones of a guesser who is convinced that he guesses right, "Guinea-pigs!"

It is unfortunately impossible to convey to the reader an adequate idea, by mere description, of the extraordinary gravity of manner, the looks of surprise and the tones of conviction which accompanied these various popular conjectures as to our calling and station in life, and which added immeasurably at the time to their comic effect. Curiously enough, whenever they took the form of questions, any jesting in returning an answer never seemed either to be appreciated or understood by the country people. Serious replies shared much the same fate as jokes. Everybody asked whether we could pay for riding, and nobody believed that we preferred walking, if we could. So we soon gave up the idea of affording any information at all; and walked through the country comfortably as mappers, trodgers, tradesmen, guinea-pig-mongers, and poor back-burdened vagabond lads, altogether, or one at a time, just as the peasantry pleased.

I have not communicated to the reader all the conjectures formed about us, for the simple reason that many of them, when they ran to any length, were by no means so intelligible as could be desired. It will readily be imagined, that in a county which had a language of its own (something similar to the Welsh) down to the time of Edward VI., if not later —in a county where this language continued to be spoken among the humbler classes until nearly the end of the seventeenth century, and where it still gives their names to men, places, and implements—some remnants of it must attach themselves to the dialect of English now spoken by the lower orders. This is enough of itself to render Cornish talk not very easy to be understood by ordinary strangers; but the difficulty of comprehending it is still further increased by the manner in which the people speak. They pronounce rapidly and indistinctly, often running separate syllables into one another through a sentence, until the whole sounds like one long fragmentary word. To the student in philology a series of conversations with the Cornish poor would, I imagine, afford ample matter for observation of the most interesting kind. Some of their expressions have a character that is quite patriarchal. Young men, for instance, are addressed by their elders as, "my son"—everything eatable, either for man or beast, is commonly denominated, "meat."

It may be expected, before I close this hasty sketch of the Cornish people, that I should touch on the dark side of the picture—unfinished though it is—which I have endeavoured to draw. But I have nothing to communicate on the subject of offences in Cornwall, beyond a few words about "wrecking" and smuggling.

Opinions have been divided among well-informed persons, as to the truth or falsehood of those statements of travellers and historians, which impute the habitual commission of outrages and robberies on sufferers by shipwreck to the Cornish of former generations. Without entering into this question of the past, which can only be treated as a matter for discussion, I am happy, in proceeding at once to the present, to be able to state, as a matter of fact, that "wrecking" is a crime unknown in the Cornwall of our day. So far from maltreating shipwrecked persons, the inhabitants of the sea-shore risk their lives to save them. I make this assertion, on the authority of a gentleman whose life has been passed in the West of Cornwall; whose avocations take him much among the poor of all ranks and characters; and who has himself seen wrecked sailors rescued from death by the courage and humanity of the population of the coast.

In reference to smuggling, many years have passed without one of those fatal encounters between smugglers and revenue officers which, in other days, gave a dark and fearful character to the contraband trade in Cornwall. So well is the coast watched, that no smuggling of any consequence can now take place. It is only the oldest Cornish men who can give you any account, from personal experience, of adventures in "running a cargo;" and those that I heard described were by no means of the romantic or interesting order.

Beyond this, I have nothing further to relate regarding criminal matters. It may not unreasonably be doubted whether a subject so serious and so extensive as the Statistics of Crime, is not out of the scope of a book like the present, whose only object is to tell a simple fireside story which may amuse an idle, or solace a mournful hour. Moreover, remembering the assistance and the kindness that my companion and I met with throughout Cornwall —and those only who have travelled on foot can appreciate how much the enjoyment of exploring a country may be heightened or decreased, according to the welcome given to the stranger by the inhabitants—remembering, too, that we walked late at night, through districts inhabited only by the roughest and poorest classes, entirely unmolested; and that we trusted much on many occasions to the honesty of the people, and never found cause to repent our trust—I cannot but feel that it would be an ungracious act to ransack newspapers and Reports to furnish materials for recording in detail, the vices of a population whom I have only personally known by their virtues. Let you and I, reader, leave off with the same pleasant impressions of the Cornish people—you, whose only object is to hear, and I whose only object is to tell, the story of a holiday walk. There is enough to be found in them that is good, amply to justify a little inattention to whatever we may discover that is bad.

V.

LOO-POOL.

"Now, I think it very much amiss," remarks Sterne, in 'Tristram Shandy,' "that a man cannot go quietly through a town and let it alone, when it does not meddle with him, but that he must be turning about, and drawing his pen at every kennel he crosses over, merely, o' my conscience, for the sake of drawing it." I quote this wise and witty observation on a bad practice of some travel-writers, as containing the best reason that I can give the reader for transporting him at once over some sixty miles of Cornish high-roads and footpaths, without stopping to drop one word of description by the way. Having left off the record of our travels at Liskeard, and taking it up again—as I mean to do here—at Helston, I skip over five intermediate market-towns and two large villages, with a mere dash of the pen. Lostwithiel, Fowey, St. Austell, Grampound, Probus, Truro, Falmouth, are all places of mark and note, and have all certain curiosities and sights of their own to interest the inquisitive tourist; but, nevertheless, not one of them "meddled" with me in the course of my rambles, and acting on Sterne's excellent principle, I purpose "letting them alone" now. In other words, the several towns and villages that I have enumerated, though presenting much that was generally picturesque and attractive in the way of old buildings and pretty scenery, exhibited little that was distinctive or original in character; produced therefore rather pleasant than vivid impressions; and would by no means suggest any very original series of descriptions to fill the pages of a book which is confined to such subjects only as are most exclusively and strikingly Cornish.

The town of Helston, where we now halt for the first time since we left the Cheese-Wring and St. Cleer's Well, might, if tested by its own merits alone, be passed over as unceremoniously as the towns already passed over before it. Its principal recommendation, in the opinion of the inhabitants, appeared to be that it was the residence of several very "genteel families," who have certainly not communicated much of their gentility to the lower orders of the population—a riotous and drunken set, the only bad specimens of Cornish people that I met with in Cornwall. The streets of Helston are a trifle larger and a trifle duller than the streets of Liskeard; the church is comparatively modern in date, and superlatively ugly in design. A miserable altar-piece, daubed in gaudy colours on the window above the communion-table, is the only approach to any attempt at embellishment in the interior. In short, the town has nothing to offer to attract the stranger, but a public festival—a sort of barbarous carnival—held there annually on the 8th of May. This festival is said to be of very ancient origin, and is called "The Furry"—an old Cornish word, signifying a gathering; and, at Helston particularly, a gathering in celebration of the return of spring. The Furry begins early in the morning with singing, to an accompaniment of drums and kettles. All the people in the town immediately leave off work and scamper into the country; having reached which, they scamper back again, garlanded with leaves and flowers, and caper about hand-in-hand through the streets, and in and out of all the houses, without let or hindrance. Even the "genteel" resident families allow themselves to be infected with the general madness, and wind up the day's capering consistently enough by a night's capering at a grand ball. A full account of these extraordinary absurdities may be found in Polwhele's "History of Cornwall."

But, though thus uninteresting in itself, Helston must be visited by every tourist in Cornwall for the sake of the grand, the almost unrivalled scenery to be met with near it. The town is not only the best starting-point from which to explore the noble line of coast rocks which ends at the Lizard Head; but possesses the further recommendation of lying in the immediate vicinity of the largest lake in Cornwall—Loo Pool.

The banks of Loo Pool stretch on either side to the length of two miles; the lake, which in summer occupies little more than half the space that it covers in winter, is formed by the flow of two or three small streams. You first reach it from Helston, after a walk of half a mile; and then see before you, on either hand, long ranges of hills rising gently from the water's edge, covered with clustering trees, or occupied by wide cornfields and sloping tracts of common land. So far, the scenery around Loo Pool resembles the scenery around other lakes; but as you proceed, the view changes in the most striking and extraordinary manner. Walking on along the winding banks of the pool, you taste the water and find it soft and fresh, you see ducks swimming about in it from the neighbouring farm-houses, you watch the rising of the fine trout for which it is celebrated—every object tends to convince you that you are wandering by the shores of an inland lake—when suddenly at a turn in the hill slope, you are startled by the shrill cry of the gull, and the fierce roar of breakers thunders on your ear—you look over the light grey waters of the lake, and behold, stretching immediately above and beyond them, the expanse of the deep blue ocean, from which they are only separated by a strip of smooth white sand!

You hurry on, and reach this bar of sand which parts the great English Channel and the little Loo Pool—a child might run across it in a minute! You stand in the centre. On one side, close at hand, water is dancing beneath the breeze in glassy, tiny ripples; on the other, equally close, water rolls in mighty waves, precipitated on the ground in dashing, hissing, writhing floods of the whitest foam—here, children are floating mimic boats on a mimic sea; there, the stateliest ships of England are sailing over the great deep—both scenes visible in one view. Rocky cliffs and arid sands appear in close combination with rounded fertile hills, and long grassy slopes; salt spray leaping over the first, spring-water lying calm beneath the last! No fairy vision of Nature that ever was imagined is more fantastic, or more lovely than this glorious reality, which brings all the most widely contrasted characteristics of a sea view and an inland view into the closest contact, and presents them in one harmonious picture to the eye.

The ridge of sand between Loo Pool and the sea, which, by impeding the flow of the inland streams spreads them in the form of a lake over the valley-ground between two hills, is formed by the action of storms from the south-west. Such, at least, is the modern explanation of the manner in which Loo Bar has been heaped up. But there is an ancient legend in connexion with it, which tells a widely different story.

It is said that the terrible Cornish giant, or ogre, Tregeagle, was trudging homewards one day, carrying a huge sack of sand on his back, which—being a giant of neat and cleanly habits—he designed should serve him for sprinkling his parlour floor. As he was passing along the top of the hills which now overlook Loo Pool, he heard a sound of scampering footsteps behind him; and, turning round, saw that he was hotly pursued by no less a person than the devil himself. Big as he was, Tregeagle lost heart and ignominiously took to his heels: but the devil ran nimbly, ran steadily, ran without losing breath—ran, in short, *like* the devil. Tregeagle was fat, short-winded, had a load on his back, and lost ground at every step. At last, just as he reached the seaward extremity of the hills, he determined in despair to lighten himself of his burden, and thus to seize the only chance of escaping his enemy by superior fleetness of foot. Accordingly, he opened his huge sack in a great hurry, shook out all his sand over the precipice, between the sea and the river which then ran into it, and so formed in a moment the Bar of Loo Pool.

In the winter time, the lake is the cause and the scene of an extraordinary ceremony. The heavy incessant rains which then fall (ice is almost unknown in the moist climate of Cornwall), increase day by day the waters of the Pool, until they encroach over the whole of the low flat valley between Helston and the sea. Then, the smooth paths of turf, the little streams that run by their side—so pleasant to look on in the summer time—are hidden by the great overflow. Mill-wheels are stopped; cottages built on the declivities of the hill are threatened with inundation. Out on the bar, at high tide, but two or three feet of sand appear between the stormy sea on the one hand, and the stagnant swollen lake on the other. If Loo Pool were measured now, it would be found to extend to a circumference of seven miles.

When the flooding of the lake has reached its climax, the millers, who are the principal sufferers by the overflow, prepare to cut a passage through the Bar for the superabundant waters of the Pool. Before they can do this, however, they must conform to a curious old custom which has been practised for centuries, and is retained down to the present day. Procuring two stout leathern purses, they tie up three halfpence in each, and then set off with them in a body to the Lord of the Manor. Presenting him with their purses, they state their case with all due formality, and request permission to cut their trench through the sand. In consideration of the threepenny recognition of his rights, the Lord of the Manor graciously accedes to the petition; and the millers, armed with their spades and shovels, start for the Bar.

Their projected labour is of the slightest kind. A mere ditch suffices to establish the desired communication: and the water does the rest for itself. On one occasion, so high was the tide on one side, and so full the lake on the other, that a man actually scraped away sand enough with his stick, to give vent to the waters of the Pool. Thus, after no very hard work, the millers achieve their object; and the spectators watching on the hill, behold a startling and magnificent scene.

Tearing away the sand on either side, floods of fresh water rush out furiously against floods of salt water leaping in, upheaved into mighty waves by the winter gale. A foaming roaring battle between two opposing forces of the same element takes place. The noise is terrific—it is heard like thunder, at great distances off. At last, the heavy, smooth, continuous flow of the fresh water prevails even over the power of the ocean. Farther and farther out, rushing through a wider and wider channel every minute, pour the great floods from the land, until the salt water is stained with an ochre colour, over a surface of twenty miles. But their force is soon spent: soon, the lake sinks lower and lower away from the slope of the hills. Then, with the high tide, the sea reappears triumphantly, dashing and leaping, in clouds of spray, through the channel in the sand—making the waters of the Pool brackish—now, threatening to swell them anew to overflowing—and now, at the ebb, leaving them to empty themselves again, in the manner of a great tidal river. No new change takes place, until a storm from the south-west comes on; and then, fresh masses of sand and shingle are forced up—the channel is refilled—the bar is reconstructed as if by a miracle. Again, the scene resumes its old features—again, there is a sea on one side, and a lake on the other. But now, the Pool occupies only its ordinary limits—now, the mill-wheels turn busily once more, and the smooth paths and gliding streams reappear in their former beauty, until the next winter rains shall come round, and the next winter floods shall submerge them again.

At the time when I visited the lake, its waters were unusually low. Here, they ran calm and shallow, into little, glassy, flowery creeks, that looked like fairies' bathing places. There, out in the middle, they hardly afforded depth enough for a duck to swim in. Near to the Bar, however, they spread forth wider and deeper; finely contrasted, in their dun colour and perfect repose, with the flashing foaming breakers on the other side. The surf forbade all hope of swimming; but, standing where the spent waves ran up deepest, and where the spray flew highest before the wind, I could take a natural shower-bath from the sea, in one direction; and the next moment, turning round in the other, could wash the sand off my feet luxuriously in the soft, fresh waters of Loo Pool.

VI.

THE LIZARD.

We had waited throughout one long rainy day at Helston—" remote, unfriended, melancholy, slow"—for a chance of finer weather before we started to explore the Lizard promontory. But our patience availed us little. The next morning, there was the soft, thick, misty Cornish rain still falling, just as it had already fallen without cessation for twenty-four hours. To wait longer, in perfect inactivity, and in the dullest of towns—doubtful whether the sky would clear even in a week's time—was beyond mortal endurance. We shouldered our knapsacks, and started for the Lizard in defiance of rain, and in defiance of our landlady's reiterated assertions that we should lose our way in the mist, when we walked inland; and should slip into invisible holes, and fall over fog-veiled precipices among the rocks, if we ventured to approach the coast.

What sort of scenery we walked through, I am unable to say. The rain was above—the mud was below—the mist was all around us. The few objects, near at hand, that we did now and then see, dripped with wet, and had a shadowy visionary look. Sometimes, we met a forlorn cow steaming composedly by the roadside—or an old horse, standing up to his fetlocks in mire, and sneezing vociferously—or a good-humoured peasant, who directed us on our road, and informed us with a grin, that this sort of "fine rain" often lasted for a fortnight. Sometimes we passed little villages built in damp holes, where trees, cottages, women scampering backwards and forwards peevishly on domestic errands, big boys with empty sacks over their heads and shoulders, gossiping gloomily against barn walls, and ill-conditioned pigs grunting for admission at closed kitchen doors, all looked soaked through and through together. Nothing, in short, could be more dreary and comfortless than our walk for the first two hours. But, after that, as we approached "Lizard Town," the clouds began to part to seaward; layer after layer of mist drove past us, rolling before the wind; peeps of faint greenish-blue sky appeared and enlarged apace. By the time we had arrived at our destination, a white, watery sunlight was falling over the wet landscape. The prognostications of our Cornish friends were pleasantly falsified. A fine day was in store for us after all.

The man who first distinguished the little group of cottages that we now looked on, by the denomination of Lizard *Town*, must have possessed magnificent ideas indeed on the subject of nomenclature. If the place looked like anything in the world, it looked like a large collection of farm out-buildings without a farm-house. Muddy little lanes intersecting each other at every possible angle; rickety little cottages turned about to all the points of the compass; ducks, geese, cocks, hens, pigs, cows, horses, dunghills, puddles, sheds, peat-stacks, timber, nets, seemed to be all indiscriminately huddled together where there was little or no room for them. To find the inn amid this confusion of animate and inanimate objects, was no easy matter; and when we at length discovered it, pushed our way through the live stock in the garden, and opened the kitchen door, this was the scene which burst instantaneously on our view:—

We beheld a small room literally full of babies, and babies' mothers. Interesting infants of the tenderest possible age, draped in long clothes and short clothes, and shawls and blankets, met the eye wherever it turned. We saw babies propped up uncomfortably on the dresser, babies rocking snugly in wicker cradles, babies stretched out flat on their backs on women's knees, babies prone on the floor toasting before a slow fire. Every one of these Cornish cherubs was crying in every variety of vocal key. Every one of their affectionate parents was talking at the top of her voice. Every one of their little elder brothers was screaming, squabbling, and tumbling down in the passage with prodigious energy and spirit. The mothers of England—and they only—can imagine the deafening and composite character of the noise which this large family party produced. To describe it is impossible.

Ere long, while we looked on it, the domestic scene began to change. Even as porters, policemen, and workmen of all sorts, gathered together on the line of rails at a station, move aside quickly and with one accord out of the way of the heavy engine slowly starting on its journey — so did the congregated mothers in the inn kitchen now move back on either hand with their babies, and clear a path for the great bulk of the hostess leisurely advancing from the fireside, to greet us at the door. From this most corpulent and complaisant of women, we received a hearty welcome, and a full explanation of the family orgies that were taking place under her roof. The great public meeting of all the babies in Lizard Town and the neighbouring villages, on which we had intruded, had been convened by the local doctor, who had got down from London, what the landlady termed a "lot of fine fresh matter," and was now about to strike a decisive blow at the small-pox, by vaccinating all the babies he could lay his hands on, at "one fell swoop." The surgical ceremonies were expected to begin in a few minutes.

This last piece of information sent us out of the house without a moment's delay. The sunlight had brightened gloriously since we had last beheld it—the rain was over—the mist was gone. But a short distance before us, rose the cliffs at the Lizard Head—the southernmost land in England—and to this point we now hastened, as the fittest spot from which to start on our rambles along the coast.

On our way thither, short as it was, we observed a novelty. In the South and West of Cornwall, the footpaths, instead of leading through or round the fields, are all on the top of the thick stone walls—some four feet high—which divide them. This curious arrangement for walking gives a startling and picturesque character to the figures of the country people, when you see them at a distance, striding along, not on the earth but above it, and often relieved throughout the whole length of their bodies against the sky. Preserving our equilibrium, on these elevated pathways, with some difficulty against the strong south-west wind that was now blowing in our faces, we soon reached the topmost rocks that crown the Lizard Head: and then, the whole noble line of coast and the wild stormy ocean opened grandly into view.

On each side of us, precipice over precipice, cavern within cavern, rose the great cliffs protecting the land against the raging sea. Three hundred feet beneath, the foam was boiling far out over a reef of black rocks. Above and around, flocks of sea-birds flew in ever lengthening circles, or perched flapping their wings and sunning their plumage, on ledges of riven stone below us. Every object forming the wide sweep of the view was on the vastest and most majestic scale. The wild varieties of form in the jagged line of rocks stretched away eastward and westward, as far as the eye could reach; black shapeless masses of mist scowled over the whole landward horizon; the bright blue sky at the opposite point was covered with towering white clouds which moved and changed magnificently; the tossing and raging of the great bright sea was sublimely contrasted by the solitude and tranquillity of the desert, overshadowed land—while ever and ever, sounding as they first sounded when the morning stars sang together, the rolling waves and the rushing wind pealed out their primeval music over the whole scene!

And now, when we began to examine the coast more in detail, inquiring the names of remarkable objects as we proceeded, we found ourselves in a country where each succeeding spot that the traveller visited, was memorable for some mighty convulsion of Nature, or tragically associated with some gloomy story of shipwreck and death. Turning from the Lizard Head towards a cliff at some little distance, we passed through a field on our way, overgrown with sweet-smelling wild flowers, and broken up into low grassy mounds. This place is called "Pistol Meadow," and is connected with a terrible event which is still spoken of by the country people with superstitious awe.

Some hundred years since, a transport-ship, filled with troops, was wrecked on the reef off the Lizard Head. Two men only were washed ashore alive.

Out of the fearful number that perished, two hundred corpses were driven up on the beach below Pistol Meadow; and there they were buried by tens and twenties together in great pits, the position of which is still revealed by the low irregular mounds that chequer the surface of the field. The place was named, in remembrance of the quantity of fire-arms, —especially pistols—found about the wreck of the ill-fated ship, at low tide, on the reef below the cliffs. To this day, the peasantry continue to regard Pistol Meadow with feelings of awe and horror, and fear to walk near the graves of the drowned men at night. Nor have many of the inhabitants yet forgotten a revolting circumstance connected by traditional report with the burial of the corpses after the shipwreck. It is said, that when dead bodies were first washed ashore, troops of ferocious, half-starved dogs suddenly appeared from the surrounding country, and could with difficulty be driven from preying on the mangled remains that were cast up on the beach. Ever since that period, the peasantry have been reported as holding the dog in abhorrence. Whether this be true or not, it is certainly a rare adventure to meet with a dog in the Lizard district. You may walk through farm-yard after farm-yard, you may enter cottage after cottage, and never hear any barking at your heels;—you may pass, on the road, labourer after labourer, and yet never find one of them accompanied, as in other parts of the country, by his favourite attendant cur.

Leaving Pistol Meadow, after gathering a few of the wild herbs growing fragrant and plentiful over the graves of the dead, we turned our steps towards the Lizard Lighthouse. As we passed before the front of the large and massive building, our progress was suddenly and startlingly checked by a hideous chasm in the cliff, sunk to a perpendicular depth of seventy feet, and measuring more than a hundred in circumference. Nothing prepares the stranger for this great gulf; no railing is placed about it; it lies hidden by rising land, and the earth all around is treacherously smooth. The first moment when you see it, is the moment when you start back instinctively from its edge, doubtful whether the hole has not yawned open in that very instant before your feet.

This chasm — melodramatically entitled by the people, "The Lion's Den"—was formed in an extraordinary manner, not many years since. In the evening the whole surface of the down above the cliff was smooth to the eye, and firm to the foot—in the morning it had opened into an enormous hole. The men who kept watch at the Lighthouse, heard no sounds beyond the moaning of the sea—felt no shock—looked out on the night, and saw that all was apparently still and quiet. Nature suffered her convulsion and effected her change in silence. Hundreds on hundreds of tons of soil had sunk down into depths beneath them, none knew in how long, or how short a time; but there the Lion's Den was in the morning, where the firm earth had been the evening before.

The explanation of the manner in which this curious landslip occurred, is to be found by descending the face of the cliff, beyond the Lion's Den, and entering a cavern in the rocks, called "Daw's Hugo" (or Cave). The place is only accessible at low water. Passing from the beach through the opening of the cavern, you find yourself in a lofty, tortuous recess, into the farthest extremity of which, a stream of light pours down from some eighty or a hundred feet above. This light is admitted through the Lion's Den, and thus explains by itself the nature of the accident by which that chasm was formed. Here, the weight of the upper soil broke through the roof of the cave; and the earth which then fell into it, was subsequently washed away by the sea, which fills Daw's Hugo at every flow of the tide. It has lately been noticed that the loose particles of ground at the bottom of the Lion's Den, still continue to sink gradually through the narrow, slanting passage into the cave already formed; and it is expected that in no very long time the lower extremity of the chasm will widen so far, as to make the sea plainly visible through it from above. At present, the effect of the two streams of light pouring into Daw's Hugo from two opposite directions—one from the Lion's Den, the other from the seaward opening in the rocks—and falling together, in cross directions on the black rugged walls of the cave and the beautiful marine ferns growing from them, is supernaturally striking and grand. Here, Rembrandt would have loved to study; for here, even *his* sublime perception of the poetry of light and shade might have received a new impulse, and learned from the teaching of Nature one immortal lesson more.

Daw's Hugo and the Lion's Den may be fairly taken as characteristic types of the whole coast scenery about the Lizard Head, in its general aspects.

Great caves and greater landslips are to be seen both eastward and westward. In calm weather you may behold the long prospects of riven rock, in their finest combination, from a boat. At such times, you may row into vast caverns, always filled by the sea, and only to be approached when the waves ripple as calmly as the waters of a lake. Then, you may see the naturally arched roof high above you, adorned in the loveliest manner by marine plants waving to and fro gently in the wind. Rocky walls are at each side of you, variegated in dark red and dark green colours —now advancing, now receding, now winding in and out, now rising straight and lofty, until their termination is hid in a pitch-dark obscurity which no man has ever ventured to fathom to its end. Beneath, is the emerald-green sea, so still and clear that you can behold the white sand far below, and can watch the fish gliding swiftly and stealthily out and in: while, all around, thin drops of moisture are dripping from above, like rain, into the deep quiet water below, with a monotonous echoing sound which half oppresses and half soothes the ear, at the same time.

On stormy days your course is different. Then, you wander along the summits of the cliffs; and looking down, through the hedges of tamarisk and myrtle that skirt the ends of the fields, see the rocks suddenly broken away beneath you into an immense shelving amphitheatre, on the floor of which the sea boils in fury, rushing through natural archways and narrow rifts. Beyond them, at intervals as the waves fall, you catch glimpses of the brilliant blue main ocean, and the outer reefs stretching into it. Often, such wild views as these are relieved from monotony, by the prospect of smooth corn-fields and pasture-lands, or by pretty little fishing villages perched among the rocks—each with its small group of boats drawn up on a slip of sandy beach, and its modest, tiny gardens rising one above another, wherever the slope is gentle, and the cliff beyond rises high to shelter them from the winter winds.

But the place at which the coast scenery of the Lizard district arrives at its climax of grandeur is Kynance Cove. Here, such gigantic specimens are to be seen of the most beautiful of all varieties of rock—the " serpentine"—as are unrivalled in Cornwall; perhaps, unrivalled anywhere. A walk of two miles along the westward cliffs from Lizard Town, brought us to the top of a precipice of three hundred feet. Looking forward from this, we saw the white sand of Kynance Cove stretching out in a half circle into the sea.

What a scene was now presented to us! It was a perfect palace of rocks! Some rose perpendicularly and separate from each other, in the shapes of pyramids and steeples—some were overhanging at the top and pierced with dark caverns at the bottom—some were stretched horizontally on the sand, here studded with pools of water, there broken into natural archways. No one of these rocks resembled another in shape, size, or position—and all, at the moment when we looked on them, were wrapped in the solemn obscurity of a deep mist; a mist which shadowed without concealing them, which exaggerated their size, and, hiding all the cliffs beyond, presented them sublimely as separate and solitary objects in the sea-view.

It was now necessary, however, to occupy as little time as possible in contemplating Kynance Cove from a distance; for if we desired to explore it, immediate advantage was to be taken of the state of the tide, which was already rapidly ebbing. Hurriedly descending the cliffs, therefore, we soon reached the sand: and here, leaving my companion to sketch, I set forth to wander among the rocks, doubtful whither to turn my steps first. While still hesitating, I was fortunate enough to meet with a guide, whose intelligence and skill well deserve such record as I can give of them here; for, to the former I was indebted for much local information and anecdote, and to the latter, for quitting Kynance Cove with all my limbs in as sound a condition as when I first approached it.

The guide introduced himself to me by propounding a sort of stranger's catechism. 1st. "Did I want to see everything?"—"Certainly." 2nd. "Was I giddy on the tops of high places?"—"No." 3rd. "Would I be so good, if I got into a difficulty anywhere, as to take it easy, and catch hold of him tight?"—"Yes, very tight!" With these answers the guide appeared to be satisfied. He gave his hat a smart knock with one hand, to fix it on his head; and pointing upwards with the other, said, "We'll try that rock first, to look into the gulls' nests, and get some wild asparagus." And away we went accordingly.

We mount the side of an immense rock which projects far out into the sea, and is the largest of the surrounding group. It is called Asparagus Island, from the quantity of wild asparagus growing among the long grass on its summit. Half way up, we cross an ugly chasm. The guide points to a small chink or crevice, barely discernible in one side of it, and says "Devil's Bellows!" Then, first courteously putting my toes for me into a comfortable little hole in the perpendicular rock side, which just fits them, he proceeds to explain himself. Through the base of the opposite extremity of the island there is a natural channel, into which the sea rushes furiously at high tide: and finding no other vent but the little crevice we now look down on, is expelled through it in long, thin jets of spray, with a roaring noise resembling the sound of a gigantic bellows at work. But the sea is not yet high enough to exhibit this phenomenon, so the guide takes my toes out of the hole again for me, just as politely as he put them in; and forthwith leads the way up higher still—expounding as he goes, the whole art and mystery of climbing, which he condenses into this axiom:—"Never loose one hand, till you've got a grip with the other; and never scramble your toes about, where toes have no business to be."

At last we reach the topmost ridge of the island, and look down upon the white restless water far beneath, and peep into one or two deserted gulls' nests, and gather wild asparagus—which I can only describe as bearing no resemblance at all, that I could discover, to the garden species. Then, the guide points to another perpendicular rock, farther out at sea, looming dark and phantom-like in the mist, and tells me that he was the man who built the cairn of stones on its top: and then he proposes that we shall go to the opposite extremity of the ridge on which we stand, and look down into "The Devil's Throat."

This desirable journey is accomplished with the greatest ease on his part, and with considerable difficulty and delay on mine—for the wind blows fiercely over us on the height; our rock track is narrow, rugged, and slippery; the sea roars bewilderingly below; and a single false step would not be attended with agreeable consequences. Soon, however, we begin to descend a little from our "bad eminence," and come to a halt before a wide, tunnelled opening, slanting sharply downwards in the very middle of the island—a black, gaping hole, into the bottom of which the sea is driven through some unknown subterranean channel, roaring and thundering with a fearful noise, which rises in hollow echoes through the aptly-named "Devil's Throat." About this hole no grass grew: the rocks rose wild, jagged, and precipitous, all around it. If ever the ghastly imagery of Dante's terrible "Vision" was realized on earth, it was realized here.

At this place, close to the mouth of the hole, the guide suggests that we shall sit down and have a little talk !—and very impressive talk it is, when he begins the conversation by bawling into my ear (and down the Devil's Throat at the same time) to make himself heard above the fierce roaring beneath us. Now, his tale is of tremendous jets of water which he has seen, during the storms of winter, shot out of the hole before which we sit, into the creek of the sea below—now, he tells me of a shipwreck off Asparagus Island, of half-drowned sailors floating ashore on pieces of timber, and dashed out to sea again just as they touched the strand, by a jet from the Devil's Throat—now, he points away in the opposite direction, under one of the steeple-shaped rocks, and speaks of a chase after smugglers that began from this place; a desperate chase, in which some of the smugglers' cargo, but not one of the smugglers themselves, was seized—now, he talks of another great hole in the landward rocks, where the sea may be seen boiling within : a hole into which a man who was fishing for fragments of a wreck fell and was drowned ; his body being sucked away through some invisible channel, never to be seen again by mortal eyes.

Anon, the guide's talk changes from tragedy to

comedy. He begins to recount odd adventures of his own with strangers. He tells me of a huge fat woman who was got up to the top of Asparagus Island, by the easiest path, and by the exertions of several guides; who, left to herself, gasped, reeled, and fell down immediately; and was just rolling off, with all the momentum of sixteen stone, over the precipice below her, when she was adroitly caught, and anchored fast to the ground, by the ankle of one leg and the calf of the other. Then he speaks of an elderly gentleman, who, while descending the rocks with him, suddenly stopped short at the most dangerous point, giddy and panic-stricken, pouring forth death-bed confessions of all his sins, and wildly refusing to move another inch in any direction. Even this man the guide got down in safety at last, by making stepping places of his hands, on which the elderly gentleman lowered himself as on a ladder, ejaculating incoherently all the way, and trembling in great agony long after he had been safely landed on the sands.

This last story ended, it is settled that we shall descend again to the beach. Stimulated by the ease with which my worthy leader goes down beneath me, I get over-confident in my dexterity, and begin to slip here, and slide there, and come to awkward pauses at precipitous places, in what would be rather an alarming manner, but for the potent presence of the guide, who is always beneath me, ready to be fallen upon. Sometimes, when I am holding on with all the necessary tenacity of grip, as regards my hands, but, " scrambling my toes about" in a very disorderly and unworkmanlike fashion, he pops his head up from below for me to sit on ; and puts my feet into crevices for me, with many apologies for taking the liberty ! Sometimes, I fancy myself treading on what feels like soft turf ; I look down, and find that I am standing like an acrobat on his shoulders, and hear him civilly entreating me to take hold of his jacket next, and let myself down over his body to the ledge where he is waiting for me. He never makes a false step, never stumbles, scrambles, hesitates, or fails to have a hand always at my service. The nautical metaphor of "holding on by your eyelids" becomes a fact in his case. He really views his employer, as porters are expected to view a package labelled "*glass with care.*" I am firmly persuaded that he could take a drunken man up and down Asparagus Island, without the slightest risk either to himself or his charge ; and I hold him in no small admiration, when, after landing on the sand with something between a tumble and a jump, I find him raising me to my perpendicular almost before I have touched the ground, and politely hoping that I feel quite satisfied, hitherto, with his conduct as a guide.

We now go across the beach to explore some caves —dry at low water—on the opposite side. Some of these are wide, lofty, and well-lighted from without. We walk in and out and around them, as if in great, irregular, Gothic halls. Some are narrow and dark. Now, we crawl into them on hands and knees; now, we wriggle onward a few feet, serpent-like, flat on our bellies; now, we are suddenly able to stand upright in pitch darkness, hearing faint moaning sounds of pent-up winds, when we are silent, and long reverberations of our own voices, when we speak. Then, as we turn and crawl out again, we soon see before us one bright speck of light that may be fancied miles and miles away—a star shining in the earth—a diamond sparkling in the bosom of the rock. This guides us out again pleasantly; and, on gaining the open air, we find that while we have been groping in the darkness, a change has been taking place in the regions of light, which has altered and is still altering the aspect of the whole scene.

It is now two o'clock. The tide is rising fast; the sea dashes, in higher and higher waves, on the narrowing beach. Rain and mist are both gone. Overhead, the clouds are falling asunder in every direction, assuming strange momentary shapes, quaint airy resemblances of the forms of the great rocks among which we stand. Height after height along the distant cliffs dawns on us gently; great golden rays shoot down over them; far out on the ocean, the waters flash into a streak of fire; the sails of ships passing there, glitter bright; yet a moment more, and the glorious sunlight bursts out over the whole view. The sea changes soon from dull grey to bright blue, embroidered thickly with golden specks, as it rolls and rushes and dances in the wind. The sand at our feet grows brighter and purer to the eye; the sea-birds flying and swooping above us, look like flashes of white light against the blue firmament; and, most beautiful of all, the wet serpentine rocks now shine forth in full splendour beneath the sun; every one of their exquisite varieties of colour becomes plainly visible—silver grey and bright yellow, dark red, deep brown, and malachite green appear, here combined in thin intertwined streaks, there outspread in separate irregular patches—glorious ornaments of the sea-shore, fashioned by no human art!—Nature's own homemade jewellery, which the wear of centuries has failed to tarnish, and the rage of tempests has been powerless to destroy!

But the hour wanes while we stand and admire; the surf dashes nearer and nearer to our feet; soon, the sea will cover the sand, and rush swiftly into the caves where we have slowly crawled. Already the Devil's Bellows is at work—the jets of spray spout forth from it with a roar. The sea thunders louder and louder in the Devil's Throat—we must gain the cliffs while we have yet time. The guide takes his leave; my companion unwillingly closes his sketch-book; and we slowly ascend on our inland way together—looking back often and often, with no feigned regret, on all that we are leaving behind us at KYNANCE COVE.

VII.

THE PILCHARD FISHERY.

If it so happened that a stranger in Cornwall went out to take his first walk along the cliffs towards the south of the county, in the month of August, that stranger could not advance far in any direction without witnessing what would strike him as a very singular and alarming phenomenon.

He would see a man standing on the extreme edge of a precipice, just over the sea, gesticulating in a very remarkable manner, with a bush in his hand; waving it to the right and the left, brandishing it over his head, sweeping it past his feet—in short, apparently acting the part of a maniac of the most dangerous character. It would add considerably to the startling effect of this sight on the stranger, if he were told, while beholding it, that the insane individual before him was paid for flourishing the bush at the rate of a guinea a week. And if he, thereupon, advanced a little to obtain a nearer view of the madman, and then observed on the sea below (as he certainly might) a well-manned boat, turning carefully to right and left exactly as the bush turned right and left, his mystification would probably be complete, and the right time would arrive to come to his rescue with a few charitable explanatory words. He would then learn that the man with the bush was an important agent in the Pilchard Fishery of Cornwall; that he had just discovered a shoal of pilchards swimming towards the land; and that the men in the boat were guided by his gesticulations alone, in securing the fish on which they and all their countrymen on the coast depend for a livelihood.

To begin, however, with the pilchards themselves, as forming one of the staple commercial commodities of Cornwall. They may be, perhaps, best described as bearing a very close resemblance to the herring, but as being rather smaller in size and having larger scales. Where they come from before they visit the Cornish coast—where those that escape the fishermen go to when they quit it, is unknown; or, at best, only vaguely conjectured.

All that is certain about them is, that they are met with, swimming past the Scilly Isles, as early as July (when they are caught with a drift-net). They then advance inland in August, during which month the principal, or "in-shore," fishing begins; visit different parts of the coast until October or November; and after that disappear until the next year. They may be sometimes caught off the south-west part of Devonshire, and are occasionally to be met with near the southernmost coast of Ireland; but beyond these two points they are never seen on any other portion of the shores of Great Britain, either before they approach Cornwall, or after they have left it.

The first sight from the cliffs of a shoal of pilchards advancing towards the land, is not a little interesting. They produce on the sea the appearance of the shadow of a dark cloud. This shadow comes on and on, until you can see the fish leaping and playing on the surface by thousands at a time, all huddled close together, and all approaching so near to the shore, that they can be always caught in some fifty or sixty feet of water. Indeed, on certain occasions, when the shoals are of considerable magnitude, the fish behind have been known to force the fish before, literally up to the beach, so that they could be taken in buckets, or even in the hand with the greatest ease. It is said that they are thus impelled to approach the land by precisely the same necessity which impels the fishermen to catch them as they appear—the necessity of getting food.

With the discovery of the first shoal, the active duties of the "look-out" on the cliffs begin. Each fishing-village places one or more of these men on the watch all round the coast. They are called "huers," a word said to be derived from the old French verb, *huer*, to call out, to give an alarm. On the vigilance and skill of the "huer" much depends. He is, therefore, not only paid his guinea a week while he is on the watch, but receives, besides, a perquisite in the shape of a per-centage on the produce of all the fish taken under his auspices. He is placed at his post, where he can command an uninterrupted view of the sea, some days before the pilchards are expected to appear; and, at the same time, boats, nets, and men are all ready for action at a moment's notice.

The principal boat used is at least fifteen tons in burden, and carries a large net called the "seine," which measures a hundred and ninety fathoms in length, and costs a hundred and seventy pounds—sometimes more. It is simply one long strip, from eleven to thirteen fathoms in breadth, composed of very small meshes, and furnished, all along its length, with lead at one side and corks at the other. The men who cast this net are called the "shooters," and receive eleven shillings and sixpence a week, and a perquisite of one basket of fish each out of every haul.

As soon as the "huer" discerns the first appearance of a shoal, he waves his bush. The signal is conveyed to the beach immediately by men and boys watching near him. The "seine" boat (accompanied by another small boat, to assist in casting the net) is rowed out where he can see it. Then there is a pause, a hush of great expectation on all sides. Meanwhile, the devoted pilchards press on—a compact mass of thousands on thousands of fish, swimming to meet their doom. All eyes are fixed on the "huer;" he stands watchful and still, until the shoal is thoroughly embayed, in water which he knows to be within the depth of the "seine" net. Then, as the fish begin to pause in their progress, and gradually crowd closer and closer together, he gives the signal; the boats come up, and the "seine" net is cast, or, in the technical phrase "shot," overboard.

The grand object is now to enclose the entire shoal. The leads sink one end of the net perpendicularly to the ground; the corks buoy up the other to the surface of the water. When it has been taken all round the fish, the two extremities are made fast, and the shoal is then imprisoned within an oblong barrier of network surrounding it on all sides. The great art is to let as few of the pilchards escape as possible, while this process is being completed. Whenever the "huer" observes from above that they are startled, and are separating at any particular point, to that point he waves his bush, thither the boats are steered, and there the net is "shot" at once. In whatever direction the fish attempt to get out to sea again, they are thus immediately met and thwarted with extraordinary readiness and skill. This labour completed, the silence of intense expectation that has hitherto prevailed among the spectators on the cliff, is broken. There is a great shout of joy on all sides —the shoal is secured!

The "seine" is now regarded as a great reservoir of fish. It may remain in the water a week or more. To secure it against being moved from its position in case a gale should come on, it is warped by two or three ropes to points of land in the cliff, and is, at the same time, contracted in circuit, by its opposite ends being brought together, and fastened tight over a length of several feet. While these operations are in course of performance, another boat, another set of men, and another net (different in form from the "seine") are approaching the scene of action.

This new net is called the "tuck;" it is smaller than the "seine," inside which it is now to be let down for the purpose of bringing the fish closely collected to the surface. The men who manage this net are termed "regular seiners." They receive ten shillings a week, and the same perquisite as the "shooters." Their boat is first of all rowed inside the seine-net, and laid close to the seine-boat, which remains stationary outside, and to the bows of which one rope at one end of the "tuck-net" is fastened. The "tuck" boat then slowly makes the inner circuit of the "seine," the smaller net being dropped overboard as she goes, and attached at intervals to the larger. To prevent the fish from getting between the two nets during this operation, they are frightened into the middle of the enclosure by beating the water, at proper places, with oars, and heavy stones fastened to ropes. When the "tuck" net has at length travelled round the whole circle of the "seine," and is securely fastened to the "seine" boat, at the end as it was at the beginning, everything is ready for the great event of the day, the hauling of the fish to the surface.

Now, the scene on shore and sea rises to a prodigious pitch of excitement. The merchants, to whom the boats and nets belong, and by whom the men are employed, join the "huer" on the cliff; all their friends follow them; boys shout, dogs bark madly; every little boat in the place puts off, crammed with idle spectators; old men and women hobble down to the beach to wait for the news. The noise, the bustle, and the agitation, increase every moment. Soon the shrill cheering of the boys is joined by the deep voices of the "seiners." There they stand, six or eight stalwart sunburnt fellows, ranged in a row in the "seine" boat, hauling with all their might at the "tuck" net, and roaring the regular nautical " Yo-heave-ho !" in chorus ! Higher and higher rises the net, louder and louder shout the boys and the idlers. The merchant forgets his dignity, and joins them; the "huer," so calm and collected hitherto, loses his self-possession and waves his cap triumphantly; even you and I, reader, uninitiated spectators though we are, catch the infection, and cheer away with the rest, as if our bread depended on the event of the next few minutes. "Hooray! hooray! Yo-hoy, hoy, hoy! Pull away, boys! Up she comes! Here they are! Here they are!" The water boils and eddies; the "tuck" net rises to the surface, and one teeming, convulsed mass of shining, glancing, silvery scales; one compact crowd of tens of thousands of fish, each one of which is madly endeavouring to escape, appears in an instant!

The noise before was as nothing compared with the noise now. Boats as large as barges are pulled up in hot haste all round the net; baskets are produced by dozens: the fish are dipped up in them, and shot out, like coals out of a sack, into the boats. Ere long, the men are up to their ankles in pilchards; they jump upon the rowing benches and work on, until the boats are filled with fish as full as they can hold, and the gunwales are within two or three inches of the water. Even yet, the shoal is not exhausted; the "tuck" net must be let down again and left ready for a fresh haul, while the boats are slowly propelled to the shore, where we must join them without delay.

As soon as the fish are brought to land, one set of

men, bearing capacious wooden shovels, jump in among them; and another set bring large hand-barrows close to the side of the boat, into which the pilchards are thrown with amazing rapidity. This operation proceeds without ceasing for a moment. As soon as one barrow is ready to be carried to the salting-house, another is waiting to be filled. When this labour is performed by night, which is often the case, the scene becomes doubly picturesque. The men with the shovels, standing up to their knees in pilchards, working energetically; the crowd stretching down from the salting-house, across the beach, and hemming in the boat all round; the uninterrupted succession of men hurrying backwards and forwards with their barrows, through a narrow way kept clear for them in the throng; the glare of the lanterns giving light to the workmen, and throwing red flashes on the fish as they fly incessantly from the shovels over the side of the boat—all combine together to produce such a series of striking contrasts, such a moving picture of bustle and animation, as not even the most careless of spectators could ever forget.

Having watched the progress of affairs on the shore, we next proceed to the salting-house, a quadrangular structure of granite, well-roofed in all round the sides, but open to the sky in the middle. Here, we must prepare ourselves to be bewildered by incessant confusion and noise; for here are assembled all the women and girls in the district, piling up the pilchards on layers of salt, at three-pence an hour; to which remuneration, a glass of brandy and a piece of bread and cheese are hospitably added at every sixth hour, by way of refreshment. It is a service of some little hazard to enter this place at all. There are men rushing out with empty barrows, and men rushing in with full barrows, in almost perpetual succession. However, while we are waiting for an opportunity to slip through the doorway, we may amuse ourselves by watching a very curious ceremony which is constantly in course of performance outside it.

As the filled barrows are going into the salting-house, we observe a little urchin running by the side of them, and hitting their edges with a long cane, in a constant succession of smart strokes, until they are fairly carried through the gate, when he quickly returns to perform the same office for the next series that arrive. The object of this apparently unaccountable proceeding is soon practically illustrated by a group of children, hovering about the entrance of the salting-house, who every now and then dash resolutely up to the barrows, and endeavour to seize on as many fish as they can take away at one snatch. It is understood to be their privilege to keep as many pilchards as they can get in this way by their dexterity, in spite of a liberal allowance of strokes aimed at their hands; and their adroitness richly deserves its reward. Vainly does the boy officially entrusted with the administration of the cane, strike the sides of the barrow with malignant smartness and perseverance—fish are snatched away with lightning rapidity and pickpocket neatness of hand. The hardest rap over the knuckles fails to daunt the sturdy little assailants. Howling with pain, they dash up to the next barrow that passes them, with unimpaired resolution; and often collect their ten or a dozen fish a piece, in an hour or two. No description can do justice to the "Jack-in-Office" importance of the boy with the cane, as he flourishes it about ferociously in the full enjoyment of his vested right to castigate his companions as often as he can. As an instance of the early development of the tyrannic tendencies of human nature, it is, in a philosophical point of view, quite unique.

But now, while we have a chance, while the doorway is accidentally clear for a few moments, let us enter the salting-house, and approach the noisiest and most amusing of all the scenes which the pilchard fishery presents. First of all we pass a great heap of fish lying in one recess inside the door, and an equally great heap of coarse, brownish salt lying in another. Then we advance farther, get out of the way of everybody, behind a pillar, and see a whole congregation of the fair sex screaming, talking, and—to their honour be it spoken—working at the same time, round a compact mass of pilchards which their nimble hands have already built up to a height of three feet, a breadth of more than four, and a length of twenty. Here we have every variety of the "fairer half of creation" displayed before us, ranged round an odoriferous heap of salted fish. Here we see crones of sixty and girls of sixteen; the ugly and the lean, the comely and the plump; the sour-tempered and the sweet—all squabbling, singing, jesting, lamenting, and shrieking at the very top of their very shrill voices for "more fish," and "more salt;" both of which are brought from the stores, in small buckets, by a long train of children running backwards and forwards with unceasing activity and in bewildering confusion. But, universal as the uproar is, the work never flags; the hands move as fast as the tongues; there may be no silence and no discipline, but there is also no idleness and no delay. Never was three-pence an hour more joyously or more fairly earned than it is here!

The labour is thus performed. After the stone floor has been swept clean, a thin layer of salt is spread on it, and covered with pilchards laid partly edgewise, and close together. Then another layer of salt, smoothed fine with the palm of the hand, is laid over the pilchards; and then more pilchards are placed upon that; and so on until the heap rises to four feet or more. Nothing can exceed the ease, quickness, and regularity with which this is done. Each woman works on her own small area, without reference to her neighbour; a bucketful of salt and a bucketful of fish being shot out in two little piles under her hands, for her own especial use. All proceed in their labour, however, with such equal diligence and equal skill, that no irregularities appear in the various layers when they are finished—they run as straight and smooth from one end to the other, as if they were constructed by machinery. The heap, when completed, looks like a long, solid, neatly-made mass of dirty salt; nothing being now seen of the pilchards but the extreme tips of their noses or tails, just peeping out in rows, up the sides of the pile.

Having now inspected the progress of the pilchard fishery, from the catching to the curing, we have seen all that we can personally observe of its different processes, at one opportunity. What more remains to be done, will not be completed until after an interval of several weeks. We must be content to hear about this from information given to us by others. Yonder, sitting against the outside wall of the salting-house, is an intelligent old man, too infirm now to do more than take care of the baby that he holds in his arms, while the baby's mother is earning her three-pence an hour inside. To this ancient we will address all our inquiries; and he is well qualified to answer us, for the poor old fellow has worked away all the pith and marrow of his life in the pilchard fishery.

The fish—as we learn from our old friend, who is mightily pleased to be asked for information—will remain in salt, or, as the technical expression is, " in bulk," for five or six weeks. During this period, a quantity of oil, salt, and water drips from them into wells cut in the centre of the stone floor on which they are placed. After the oil has been collected and clarified, it will sell for enough to pay off the whole expense of the wages, food, and drink given to the "seiners"—perhaps defraying other incidental charges besides. The salt and water left behind, and offal of all sorts found with it, furnish a valuable manure. Nothing in the pilchard itself, or in connexion with the pilchard, runs to waste—the precious little fish is a treasure in every part of him.

After the pilchards have been taken out of "bulk," they are washed clean in salt water, and packed in hogsheads, which are then sent for exportation to some large sea-port — Penzance for instance—in coast traders. The fish reserved for use in Cornwall, are generally cured by those who purchase them. The export trade is confined to the shores of the Mediterranean—Italy and Spain providing the two great foreign markets for pilchards. The home consumption, as regards Great Britain, is nothing, or next to nothing. Some variation takes place in the prices realized by the foreign trade— their average, wholesale, is stated to be about fifty shillings per hogshead.

As an investment for money, on a small scale, the pilchard fishery offers the first great advantage of security. The only outlay necessary, is that for providing boats and nets, and for building salting-houses—an outlay which, it is calculated, may be covered by a thousand pounds. The profits resulting from the speculation are immediate and large. Transactions are managed on the ready money principle, and the markets of Italy and Spain (where pilchards are considered a great delicacy) are always open to any supply. The fluctuation between a good season's fishing and a bad season's fishing is rarely, if ever, seriously great. Accidents happen but seldom; the casualty most dreaded, being the enclosure of a large fish along with a shoal of pilchards. A "ling," for instance, if unfortunately imprisoned in the seine, often bursts through its thin meshes, after luxuriously gorging himself with prey, and is of course at once followed out of the breach by all the pilchards. Then, not only is the shoal lost, but the net is seriously damaged, and must be tediously and expensively repaired. Such an accident as this, however, very seldom happens; and when it does, the loss occasioned falls on those best able to bear it, the merchant speculators. The work and wages of the fishermen go on as usual.

Some idea of the almost incalculable multitude of pilchards caught on the shores of Cornwall, may be formed from the following *data.* At the small fishing cove of Trereen, 600 hogsheads were taken in little more than one week, during August, 1850. Allowing 2,400 fish only to each hogshead—3,000 would be the highest calculation—we have a result of 1,440,000 pilchards, caught by the inhabitants of one little village alone, on the Cornish coast, at the commencement of the season's fishing.

At considerable sea-port towns, where there is an unusually large supply of men, boats, and nets, such figures as those quoted above, are far below the mark. At St. Ives, for example, 1,000 hogsheads were taken in the first three seine nets cast into the water. The number of hogsheads exported annually, averages 22,000. In 1850, 27,000 were secured for the foreign markets. Incredible as these numbers may appear to some readers, they may nevertheless be relied on; for they are derived from trustworthy sources—partly from local returns furnished to me; partly from the very men who filled the baskets from the boat-side, and who afterwards verified their calculations by frequent visits to the salting-houses.

Such is the pilchard fishery of Cornwall—a small unit, indeed, in the vast aggregate of England's internal sources of wealth : but yet neither unimportant nor uninteresting, if it be regarded as giving active employment to a hardy and honest race who would starve without it; as impartially extending the advantages of commerce to one of the remotest corners of our island; and, more than all, as displaying a wise and beautiful provision of Nature, by which the rich tribute of the great deep is most generously lavished on the land most in need of a compensation for its own sterility.

VIII.

THE LAND'S END.

SOMETHING like what Jerusalem was to the pilgrim in the Holy Land, the Land's End is—comparing great things with small—to the tourist in Cornwall. It is the Ultima Thule where his progress stops—the shrine towards which his face has been set, from the first day when he started on his travels—the main vent, through which all the pent-up enthusiasm accumulated along the line of route is to burst its way out, in one long flow of admiration and delight.

The Land's End! There is something in the very words that stirs us all. It was the name that struck us most, and was best remembered by us, as children, when we learnt our geography. It fills the minds of imaginative people with visions of barrenness and solitude, with dreams of some lonely promontory, far away by itself out in the sea—the sort of place where the last man in England would be most likely to be found waiting for death, at the end of the world! It suggests even to the most prosaically constituted people, ideas of tremendous storms, of flakes of foam flying over the land before the wind, of billows in convulsion, of rocks shaken to their centre, of caves where smugglers lurk in ambush, of wrecks and hurricanes, desolation, danger, and death. It awakens curiosity in the most careless —once hear of it, and you long to see it—tell your friends that you have travelled in Cornwall, and ten thousand chances to one, the first question they ask is:—" Have you been to the Land's End?"

And yet, strange to say, this spot so singled out and set apart by our imaginations as something remarkable and even unique of its kind, is as a matter of fact, not distinguishable from any part of the coast on either side of it, by any local peculiarity whatever. If you desire really and truly to stand on the Land's End itself, you must ask your way to it, or you are in danger of mistaking any one of the numerous promontories on the right hand and the left, for your actual place of destination. But I am anticipating. Before I say more about the Land's

End, it is necessary to relate how my companion and I got there, and what we saw that was interesting and characteristic on our road.

The reader may perhaps remember that he last left us scrambling out of reach of the tide, up the cliffs overlooking Kynance Cove. From that place we got back to Helston in mist and rain, just as we had left it. From Helston we proceeded to Marazion, —stopping there to visit St. Michael's Mount, so well known to readers of all classes by innumerable pictures and drawings, and by descriptions scarcely less plentiful, that they will surely be relieved rather than disappointed, if these pages exhibit the distinguished negative merit of passing the Mount without notice. From Marazion we walked to Penzance, from Penzance to the beautiful coast scenery at Lamorna Cove, and thence to Trereen, celebrated as the halting place for a visit to one of Cornwall's greatest curiosities—the Loggan Stone.

This far-famed rock rises on the top of a bold promontory of granite, jutting far out into the sea, split into the wildest forms, and towering precipitously to a height of a hundred feet. When you reach the Loggan Stone, after some little climbing up perilous-looking places, you see a solid, irregular mass of granite, which is computed to weigh eighty five tons, supported by its centre only, on a flat, broad rock, which, in its turn, rests on several others stretching out around it on all sides. You are told by the guide to turn your back to the uppermost stone; to place your shoulders under one particular part of its lower edge, which is entirely disconnected, all round, with the supporting rock below; and in this position to push upwards slowly and steadily, then to leave off again for an instant, then to push once more, and so on, until after a few moments of exertion, you feel the whole immense mass above you moving as you press against it. You redouble your efforts—then turn round—and see the massy Loggan Stone, set in motion by nothing but your own pair of shoulders, slowly rocking backwards and forwards with an alternate ascension and declension, at the outer edges, of at least three inches. You have treated eighty-five tons of granite like a child's cradle; and, like a child's cradle, those eighty-five tons have rocked at your will!

The pivot on which the Loggan Stone is thus easily moved, is a small protrusion in its base, on all sides of which the whole surrounding weight of rock is, by an accident of Nature, so exactly equalized, as to keep it poised in the nicest balance on the one little point in its lower surface which rests on the flat granite slab beneath. But perfect as this balance appears at present, it has lost something, the merest hair's-breadth, of its original faultlessness of adjustment. The rock is not to be moved now, either so easily or to so great an extent, as it could once be moved. Six-and-twenty years since, it was overthrown by artificial means; and was then lifted again into its former position. This is the story of the affair, as it was related to me by a man who was an eye-witness of the process of restoring the stone to its proper place.

In the year 1824, a certain Lieutenant in the Royal Navy, then in command of a cutter stationed off the southern coast of Cornwall, was told of an ancient Cornish prophecy, that no human power should ever succeed in overturning the Loggan Stone. No sooner was the prediction communicated to him, than he conceived a mischievous ambition to falsify practically an assertion which the commonest common sense might have informed him had sprung from nothing but popular error and popular superstition. Accompanied by a body of picked men from his crew, he ascended to the Loggan Stone, ordered several levers to be placed under it at one point, gave the word to "heave"—and the next moment had the miserable satisfaction of seeing one of the most remarkable natural curiosities in the world utterly destroyed, for aught he could foresee to the contrary, under his own directions!

But Fortune befriended the Loggan Stone. One edge of it, as it rolled over, became fixed by a lucky chance in a crevice in the rocks immediately below the granite slab from which it had been started. Had this not happened, it must have fallen over a sheer precipice, and been lost in the sea. By another accident, equally fortunate, two labouring men at work in the neighbourhood, were led by curiosity secretly to follow the Lieutenant and his myrmidons up to the Stone. Having witnessed, from a secure hiding-place, all that occurred, the two workmen, with great propriety, immediately hurried off to inform the lord of the manor of the wanton act of destruction which they had seen perpetrated.

The news was soon communicated throughout the district, and thence, throughout all Cornwall. The indignation of the whole county was aroused. Antiquaries, who believed the Loggan Stone to have been balanced by the Druids; philosophers who held that it was produced by an eccentricity of natural formation; ignorant people, who cared nothing about Druids, or natural formations, but who liked to climb up and rock the stone whenever they passed near it; tribes of guides who lived by showing it; innkeepers in the neighbourhood, to whom it had brought customers by hundreds; tourists of every degree who were on their way to see it—all joined in one general clamour of execration against the overthrower of the rock. A full report of the affair was forwarded to the Admiralty; and the Admiralty, for once, acted vigorously for the public advantage, and mercifully spared the public purse.

The Lieutenant was officially informed that his commission was in danger, unless he set up the Loggan Stone again in its proper place. The materials for compassing this achievement were offered to him, *gratis*, from the Dock Yards; but he was left to his own resources to defray the expense of employing workmen to help him. Being by this time awakened to a proper sense of the mischief he had done, and to a tolerably strong conviction of the disagreeable position in which he was placed with the Admiralty, he addressed himself vigorously to the task of repairing his fault. Strong beams were planted about the Loggan Stone, chains were passed round it, pulleys were rigged, and capstans were manned. After a week's hard work and brave perseverance on the part of every one employed in the labour, the rock was pulled back into its former position, but not into its former perfection of balance: it has never moved since as freely as it moved before.

It is only fair to the Lieutenant to add to this narrative of his mischievous frolic the fact, that he defrayed, though a poor man, all the heavy expenses of replacing the rock. Just before his death, he paid the last remaining debt, and paid it with interest.

Leaving the Loggan Stone, we next shaped our course for the Land's End. We stopped on our way, to admire the desolate pile of rocks and caverns which form the towering promontory, called "Tol-Peden-Penwith," or, "The Holed Headland on the Left." Thence, turning a little inland—passing over wild, pathless moors; occasionally catching distant glimpses of the sea, with the mist sometimes falling thick down to the very edges of the waves, sometimes parting mysteriously and discovering distant crags of granite rising shadowy out of the foaming waters,—we reached, at last, the limits of our outward journey, and saw the Atlantic before us, rolling against the westernmost extremity of the shores of England.

I have already said, that the stranger must ask his way before he can find out the particular mass of rocks, geographically entitled to the appellation of the "Land's End." He may, however, easily discover when he has reached the *district* of the "Land's End," by two rather remarkable indications that he will meet with on his road. He will observe, at some distance from the coast, an old milestone marked "I," and will be informed that this is the real original first mile in England; as if all measurement of distances began strictly from the West! A little further on he will come to a house, on one wall of which he will see written in large letters, "This is the first Inn in England," and on the other: "This is the last Inn in England;" as if the recognised beginning, and end too, of the Island of Britain were here, and here only! Having pondered a little on the slightly exclusive view of the attributes of their locality, taken by the inhabitants, he will then be led forward, about half a mile, by his guide, will descend some cliffs, will walk out on a ridge of rocks till he can go no farther—and will then be told that he is standing on the Land's End!

Here, as elsewhere, there are certain "sights" which a stranger is required to examine assiduously, as a duty if not as a pleasure, by guide-book law, rigidly administered by guides. There is, first of all, the mark of a horse's hoof, which is with great care kept *sharply modelled* (to borrow the painter's phrase), in the thin grass at the edge of a precipice. This mark commemorates the narrow escape from death of a military man who, for a wager, rode a horse down the cliff to the extreme verge of the Land's End; where the poor animal, seeing its danger, turned in affright, reared, and fell back into the sea raging over the rocks beneath. The foolhardy rider had just sense enough left to throw himself off in time—he tumbled on the ground, within a few inches of the precipice, and so barely saved the life which he had richly deserved to lose.

After the mark of the hoof, the traveller is next desired to look at a natural tunnel in the outer cliff, which pierces it through from one end to the other. Then his attention is directed to a lighthouse built on a reef of rocks detached from the land; and he is told of the great waves which break over the top of the building during the winter storms. Lastly, he is requested to inspect a quaint protuberance in a pile of granite at a little distance off, which bears a remote resemblance to a gigantic human face, adorned with a short beard; and which, he is informed, is considered quite a portrait (of all the people in the world to liken it to!) of Dr. Johnson! It is, therefore, publicly known as "Johnson's Head." If it can fairly be compared with any of the countenances of any remarkable characters that ever existed, it may be said to exhibit, in violent exaggeration, the worst physiognomical peculiarities of Nero and Henry the Eighth, combined in one face!

These several local curiosities duly examined, you are at last left free to look at the Land's End in your own way. Before you, stretches the wide, wild ocean; the largest of the Scilly Islands being barely discernible on the extreme horizon, on clear days. Tracts of heath; fields where corn is blown by the wind into mimic waves; downs, valleys, and crags, mingle together picturesquely and confusedly, until they are lost in the distance, on your left. On your right is a magnificent bay, bounded at either extremity by far-stretching promontories rising from a beach of the purest white sand, on which the yet whiter foam of the surf is ever seething, as waves on waves break one behind the other. The whole bold view possesses all the sublimity that vastness and space can bestow; but it is that sublimity which is to be seen, not described, which the heart may acknowledge and the mind contain, but which no mere words may delineate—which even painting itself may but faintly reflect.

However, it is, after all, the walk to the Land's End along the southern coast, rather than the Land's End itself, which displays the grandest combinations of scenery in which this grandest part of Cornwall abounds. There, Nature appears in her most triumphant glory and beauty—there, every mile as you proceed, offers some new prospect, or awakens some fresh impression. All objects that you meet with, great and small, moving and motionless, seem united in perfect harmony to form a scene where original images might still be found by the poet; and where original pictures are waiting, ready composed, for the painter's eye.

On approaching the wondrous landscapes between Trereen and the Land's End, the first characteristic that strikes you, is the change that has taken place in the forms of the cliffs since you left the

Lizard Head. You no longer look on variously shaped and variously coloured "serpentine" rocks; it is granite, and granite alone, that appears everywhere—granite, less lofty and less eccentric in form than the "serpentine" cliffs and crags; but presenting an appearance of adamantine solidity and strength, a mighty breadth of outline and an unbroken vastness of extent, nobly adapted to the purpose of protecting the shores of Cornwall, where they are most exposed to the fury of the Atlantic waves. In these wild districts, the sea rolls and roars in fiercer agitation, and the mists fall thicker, and at the same time fade and change faster, than elsewhere. Vessels pitching heavily in the waves, are seen to dawn, at one moment, in the clearing atmosphere—and then, at another, to fade again mysteriously, as it abruptly thickens, like phantom ships. Up on the top of the cliffs, furze and heath in brilliant clothing of purple and yellow, cluster close round great white, weird masses of rock, dotted fantastically with patches of grey-green moss. The solitude on these heights is unbroken—no houses are to be seen—often, no pathway is to be found. You go on, guided by the *sight* of the sea, when the sky brightens fitfully: and by the *sound* of the sea, when you stray instinctively from the edge of the cliff, as mist and darkness gather once more densely and solemnly all around you.

Then, when the path appears again—a winding path, that descends rapidly—you gradually enter on a new scene. Old horses startle you, scrambling into perilous situations, to pick dainty bits by the hillside; sheep, fettered by the fore and hind leg, hobble away desperately as you advance. Suddenly, you discern a small strip of beach shut in snugly between protecting rocks. A spring bubbles down from an inland valley; while not far off, an old stone well collects the water into a calm, clear pool. Sturdy little cottages, built of rough granite, and thickly thatched, stand near you, with gulls' and cormorants' eggs set in their loop-holed windows for ornament; great white sections of fish hang thickly together on their walls to dry, looking more like many legs of many dirty duck trousers, than anything else; pigsties are hard-by the cottages, either formed by the Cromlech stones of the Druids, or excavated like caves in the side of the hill. Down on the beach, where the rough old fishing-boats lie, the sand is entirely formed by countless multitudes of the tiniest, fairy-like shells, often as small as a pin's head, and all exquisitely tender in colour and wonderfully varied in form. Up the lower and flatter parts of the hills above, fishing nets are stretched to dry. While you stop to look forth over the quiet, simple scene, wild little children peep out at you in astonishment; and hard-working men and women greet you with a hearty Cornish salutation, as you pass near their cottage doors.

You walk a few hundred yards inland, up the valley, and discover in a retired, sheltered situation, the ancient village church, with its square grey tower surmounted by moss-grown turrets, with its venerable Saxon stone cross in the churchyard—where the turf graves rise humbly by twos and threes, and where the old coffin-shaped stone stands midway at the entrance gates, still used, as in former times, by the bearers of a rustic funeral. Appearing thus amid the noblest scenery, as the simple altar of the prayers of a simple race, this is a church which speaks of religion in no formal or sectarian tone. Appealing to the heart of every traveller be his creed what it may, in loving and solemn accents, it sends him on his way again, up the mighty cliffs and through the mist driving cloud-like over them, the better fitted for his journey forward here; the better fitted, it may be, even for that other dread journey of one irrevocable moment—the last he shall ever take—to his abiding-place among the spirits of the dead!

These are some of the attractions which home rambles can offer to tempt the home traveller; for these are the impressions produced, and the incidents presented during a walk to the Land's End.

IX.

BOTALLACK MINE.

I HAVE little doubt that the less patient among the readers of this narrative have already, while perusing it, asked themselves some such questions as these:—"Is not Cornwall a celebrated mineral country? Why has the author not taken us below the surface yet? Why have we heard nothing all this time about the mines?"

Readers who have questioned thus, may be assured that their impatience to go down a mine, in this book, was fully equalled by our impatience to go down a mine, in the county of which this book treats. Our anxiety, however, when we mentioned it to Cornish friends, was invariably met by the same answer. "Wait"—they all said—"until you have turned your backs on the Land's End; and then go to Botallack. The mine there is the most extraordinary mine in Cornwall; go down that, and you will not want to go down another—wait for Botallack." And we did wait for Botallack, just as the reader has waited for it in these pages. May he derive as much satisfaction from the present description of the mine, as we did from visiting the mine itself!

We left the Land's End, feeling that our homeward journey had now begun from that point; and walking northward, about five miles along the coast, arrived at Botallack. Having heard that there was some disinclination in Cornwall to allow strangers to go down the mines, we had provided ourselves—through the kindness of a friend—with a proper letter of introduction, in case of emergency. We were told to go to the counting-house to present our credentials; and on our road thither, we beheld the buildings and machinery of the mine, literally stretching down the precipitous face of the cliff, from the land at the top, to the sea at the bottom.

This sight was, in its way, as striking and extraordinary as the first view of the Cheese-Wring itself. Here, we beheld a scaffolding perched on a rock that rose out of the waves—there, a steam-pump was at work raising gallons of water from the mine every minute, on a mere ledge of land half way down the steep cliff side. Chains, pipes, conduits, protruded in all directions from the precipice; rotten-looking wooden platforms, running over deep chasms, supported great beams of timber and heavy coils of cable; crazy little boarded houses were built, where gulls' nests might have been found in other places. There did not appear to be a foot of level space anywhere, for any part of the works of the mine to stand upon; and yet, there they were, fulfilling all the purposes for which they had been constructed, as safely and completely on rocks in the sea, and down precipices in the land, as if they had been cautiously founded on the tracts of smooth solid ground above!

The counting-house was built on a projection of earth about midway between the top of the cliff and the sea. When we got there, the agent, to whom our letter was addressed, was absent; but his place was supplied by two miners who came out to receive us; and to one of them we mentioned our recommendation, and modestly hinted a wish to go down the mine forthwith.

But our new friend was not a person who did anything in a hurry. He was a grave, courteous, and rather melancholy man, of great stature and strength. He looked on us with a benevolent, paternal expression, and appeared to think that we were nothing like strong enough, or cautious enough to be trusted down the mine. "Did we know," he urged, "that it was dangerous work?" "Yes; but we didn't mind danger!"—"Perhaps we were not aware that we should perspire profusely, and be dead tired getting up and down the ladders?" "Very likely; but we didn't mind that, either!"—"Surely we shouldn't like to strip and put on miners' clothes?" "Yes, we should, of all things!" and pulling off coat and waistcoat, on the spot, we stood half-undressed already, just as the big miner was proposing another objection, which, under existing circumstances, he good-naturedly changed into a speech of acquiescence. "Very well, gentlemen," he said, taking up two suits of miners' clothes, "I see you are determined to go down; and so you shall! You'll be wet through with the heat and the work before you come up again; so just put on these things, and keep your own clothes dry."

The clothing consisted of a flannel shirt, flannel drawers, canvas trousers, and a canvas jacket—all stained of a tawny copper colour; but all quite clean A white night-cap and a round hat, composed of some iron-hard substance, well calculated to protect the head from any loose stones that might fall on it, completed the equipment; to which, three tallow-candles were afterwards added, two to hang at the button-hole, one to carry in the hand.

My friend was dressed first. He had got a suit which fitted him tolerably, and which, as far as appearances went, made a miner of him at once. Far different was my case.

The same mysterious dispensation of fate, which always awards tall wives to short men, decreed that a suit of the big miner's should be reserved for me. He stood six feet two inches—I stand five feet six inches. I put on his flannel shirt—it fell down to my toes, like a bedgown; his drawers—and they flowed in Turkish luxuriance over my feet. At his trousers I helplessly stopped short, lost in the voluminous recesses of each leg. The big miner, like a good Samaritan as he was, came to my assistance. He put the pocket button through the waist button-hole, to keep the trousers up in the first instance; then, he pulled steadily at the braces until my waistband was under my armpits; and then he pronounced that I and my trousers fitted each other in great perfection. The cuffs of the jacket were next turned up to my elbows—the white nightcap was dragged over my ears—the round hat was jammed down over my eyes. When I add to all this, that I am so near-sighted as to be obliged to wear spectacles, and that I finished my toilet by putting my spectacles on (knowing that I should see little or nothing without them), nobody, I think, will be astonished to hear that my companion seized his sketch-book, and caricatured me on the spot; and that the grave miner, polite as he was, shook with internal laughter, when I took up my tallow-candles and reported myself ready for a descent into the mine.

We left the counting-house, and ascended the face of the cliff—then, walked a short distance along the edge, descended a little again, and stopped at a wooden platform built across a deep gully. Here, the miner pulled up a trap-door, and disclosed a perpendicular ladder leading down to a black hole, like the opening of a chimney. "This is the shaft; I will go down first, to catch you in case you tumble; follow me and hold tight;" saying this, our friend squeezed himself through the trap-door, and we went after him as we had been bidden.

The black hole, when we entered it, proved to be not quite so dark as it had appeared from above.

Rays of light occasionally penetrated it through chinks in the outer rock. But by the time we had got some little way farther down, these rays began to fade. Then, just as we seemed to be lowering ourselves into total darkness, we were desired to stand on a narrow landing-place opposite the ladder, and wait there while the miner went below for a light. He soon reascended to us, bringing, not only the light he had promised, but a large lump of damp clay with it. Having lighted our candles he stuck them against the front of our hats with the clay—in order, as he said, to leave both our hands free to us to use as we liked. Thus strangely accoutred, like Solomon Eagles in the Great Plague, with flame on our heads, we resumed the descent of the shaft ; and now at last began to penetrate beneath the surface of the earth in good earnest.

The process of getting down the ladders was not very pleasant. They were all quite perpendicular the rounds were placed at irregular distances, were many of them much worn away, and were slippery with water and copper-ooze. Add to this, the narrowness of the shaft, the dripping wet rock shutting you in, as it were, all round your back and sides against the ladder—the fathomless darkness beneath

—the light flaring immediately above you, as if your head was on fire—the voice of the miner below, rumbling away in dull echoes lower and lower into the bowels of the earth—the consciousness that if the rounds of the ladder broke, you might fall down a thousand feet or so of narrow tunnel in a moment—imagine all this, and you may easily realize what are the first impressions produced by a descent into a Cornish mine.

By the time we had got down seventy fathoms, or four hundred and twenty feet of perpendicular ladders, we stopped at another landing-place, just broad enough to afford standing room for us three. Here, the miner, pointing to an opening yawning horizontally in the rock at one side of us, said that this was the first gallery from the surface; that we had done with the ladders for the present; and that a little climbing and crawling were now to begin.

Our path was a strange one, as we advanced through the rift. Rough stones of all sizes, holes here, and eminences there, impeded us at every yard. Sometimes, we could walk on in a stooping position—sometimes, we were obliged to crawl on our hands and knees. Occasionally, greater difficulties than these presented themselves. Certain parts of the gallery dipped into black, ugly-looking pits, crossed by thin planks, over which we walked dizzily, a little bewildered by the violent contrast between the flaring light that we carried above us, and the pitch darkness beneath and before us. One of these places terminated in a sudden rising in the rock, hollowed away below, but surmounted by a narrow projecting wooden platform, to which it was necessary to climb by cross-beams arranged at wide distances. My companion ascended to this awkward elevation, without hesitating; but I came to an "awful pause" before it. Fettered as I was by my Brobdignag jacket and trousers, I felt a humiliating consciousness that any extraordinary gymnastic exertion was altogether out of my power.

Our friend the miner saw my difficulty, and extricated me from it at once, with a promptitude and skill which deserve record. Descending half way by the beams, he clutched with one hand that hinder part of my too voluminous nether garments, which presented the broadest superficies of canvas to his grasp (I hope the delicate reader appreciates my ingenious indirectness of expression, when I touch on the unmentionable subject of trousers!). Grappling me thus, and supporting himself by his free hand, he lifted me up as easily as if I had been a small parcel; then carried me horizontally along the loose boards, like a refractory little boy borne off by the usher to the master's birch; or—considering the candle burning on my hat, and the necessity of elevating my position by as lofty a comparison as I can make—like a flying Mercury with a star on his head; and finally deposited me safely upon my legs again, on the firm rock pathway beyond. "You are but a light and a little man, my son," says this excellent fellow, snuffing my candle for me before we go on; "only let me lift you about as I like, and you shan't come to any harm while I am with you!"

Speaking thus, the miner leads us forward again. After we have walked a little farther in a crouching position, he calls a halt, makes a seat for us by sticking a piece of old board between the rocky walls of the gallery, and then proceeds to explain the exact subterranean position which we actually occupy.

We are now four hundred yards out, *under the bottom of the sea;* and twenty fathoms or a hundred and twenty feet below the sea level. Coast-trade vessels are sailing over our heads. Two hundred and forty feet beneath us men are at work, and there are galleries deeper yet, even below that! The extraordinary position down the face of the cliff, of the engines and other works on the surface, at Botallack, is now explained. The mine is not excavated like other mines under the land, but under the sea!

Having communicated these particulars, the miner next tells us to keep strict silence and listen. We obey him, sitting speechless and motionless. If the reader could only have beheld us now, dressed in our copper-coloured garments, huddled close together in a mere cleft of subterranean rock, with flame burning on our heads and darkness enveloping our limbs—he must certainly have imagined, without any violent stretch of fancy, that he was looking down upon a conclave of gnomes.

After listening for a few moments, a distant, unearthly noise becomes faintly audible—a long, low, mysterious moaning, which never changes, which is *felt* on the ear as well as *heard* by it—a sound that might proceed from some incalculable distance, from some far invisible height—a sound so unlike anything that is heard on the upper ground, in the free air of heaven; so sublimely mournful and still; so ghostly and impressive when listened to in the subterranean recesses of the earth, that we continue instinctively to hold our peace, as if enchanted by it, and think not of communicating to each other the awe and astonishment which it has inspired in us from the very first.

At last, the miner speaks again, and tells us that what we hear is the sound of the surf, lashing the rocks a hundred and twenty feet above us, and of the waves that are breaking on the beach beyond. The tide is now at the flow, and the sea is in no extraordinary state of agitation: so the sound is low and distant just at this period. But, when storms are at their height, when the ocean hurls mountain after mountain of water on the cliffs, then the noise is terrific; the roaring heard down here in the mine is so inexpressibly fierce and awful, that the boldest men at work are afraid to continue their labour. All ascend to the surface, to breathe the upper air and stand on the firm earth: dreading, though no such catastrophe has ever happened yet, that the sea will break in on them if they remain in the caverns below.

Hearing this, we get up to look at the rock above us. We are able to stand upright in the position we now occupy; and flaring our candles hither and thither in the darkness, can see the bright pure copper streaking the dark ceiling of the gallery in every direction. Lumps of ooze, of the most lustrous green colour, traversed by a natural net-work of thin red veins of iron, appear here and there in large irregular patches, over which water is dripping slowly and incessantly in certain places. This is the salt water percolating through invisible crannies in the rock. On stormy days it spirts out furiously in thin, continuous streams. Just over our heads we observe a wooden plug of the thickness of a man's leg; there is a hole here, and the plug is all that we have to keep out the sea.

Immense wealth of metal is contained in the roof of this gallery, throughout its whole length; but it remains, and will always remain, untouched. The miners dare not take it, for it is part, and a great part, of the rock which forms their only protection against the sea; and which has been so far worked away here, that its thickness is limited to an average of three feet only between the water and the gallery in which we now stand. No one knows what might be the consequence of another day's labour with the pickaxe on any part of it.

This information is rather startling when communicated at a depth of four hundred and twenty feet under ground. We should decidedly have preferred to receive it in the counting-house! It makes us pause for an instant, to the miner's infinite amusement, in the very act of knocking away a tiny morsel of ore from the rock, as a memento of Botallack. Having, however, ventured on reflection to assume the responsibility of weakening our defence against the sea, by the length and breadth of an inch, we secure our piece of copper, and next proceed to discuss the propriety of descending two hundred and forty feet more of ladders, for the sake of visiting that part of the mine where the men are at work.

Two or three causes concur to make us doubt the wisdom of going lower. There is a hot, moist, sickly vapour floating about us, which becomes more oppressive every moment; we are already perspiring at every pore, as we were told we should; and our hands, faces, jackets, and trousers are all more or less covered with a mixture of mud, tallow, and iron-drippings, which we can feel and smell much more acutely than is exactly desirable. We ask the miner what there is to see lower down. He replies, nothing but men breaking ore with pickaxes; the galleries of the mine are alike, however deep they may go; when you have seen one you have seen all.

The answer decides us—we determine to get back to the surface.

We returned along the gallery, just as we had advanced, with the same large allowance of scrambling, creeping, and stumbling on our way. I was charitably carried along and down the platform over the pit, by my trousers, as before; our order of procession only changing when we gained the ladders again. Then, our friend the miner went last instead of first, upon the same principle of being ready to catch us if we fell, which led him to precede us on our descent. Except that one of the rounds cracked under his weight as we went up, we ascended without casualties of any kind. As we neared the mouth of the shaft, the daylight atmosphere looked dazzlingly white, after the darkness in which we had been groping so long; and when we once more stood out on the cliff, we felt a cold, health-giving purity in the sea breeze, and, at the same time, a sense of recovered freedom in the power that we now enjoyed of running, jumping, and stretching our limbs in perfect security, and with full space for action, which it was almost a new sensation to experience. Habit teaches us to think little of the light and air that we live and breathe in, or, at most, to view them only as the ordinary conditions of our being. To find out that they are more than this, that they are a luxury as well as a necessity of life, go down into a mine, and compare what you *can* exist in there, with what you *do* exist in, on upper earth!

On re-entering the counting-house, we were greeted by the welcome appearance of two large tubs of water, with soap and flannel placed invitingly by their sides. Copious ablutions and clean clothes are potent restorers of muscular energy. These, and a half hour of repose, enabled us to resume our knapsacks as briskly as ever, and walk on fifteen miles to the town of St. Ives—our resting place for the night.

While we were sitting in the counting-house, we had some talk with our good-humoured and intelligent guide, on the subject of miners and mining at Botallack. Some of the local information that he gave us, may interest the reader—to whom I do not pretend to offer more here than a simple record of a half hour's gossip. I could only write elaborately about the Cornish mines, by swelling my pages with extracts on the subject from Encyclopædias and Itineraries which are within easy reach of every one, and on the province of which, it is neither my business nor my desire to intrude.

Botallack mine is a copper mine; but tin, and occasionally iron, are found in it as well. It is situated at the western extremity of the great strata of copper, tin, and lead, running eastward through Cornwall, as far as the Dartmoor Hills. According to the statement of my informant in the counting house, it has been worked for more than a century. In former times, it produced enormous profits to the speculators; but now the case is altered. The price of copper has fallen of late years; the lodes have proved neither so rich nor so extensive, as at past periods; and the mine, when we visited Cornwall, had failed to pay the expenses of working it.

The organization of labour at Botallack, and in all other mines throughout the county, is thus managed:—The men work eight hours underground, out of the twenty-four; taking their turn of night duty (for labour proceeds in the mines by night as well as by day), in regular rotation. The different methods on which their work is undertaken, and the rates of remuneration that they receive, have been already touched on, in the chapter on the "Cornish People." It will be found that ordinary wages for mine labour, are there stated as ranging from forty to fifty shillings a month—mention being made at the same time, of the larger remuneration which may be obtained by working "on tribute," or, in other words, by agreeing to excavate the lodes of metal for a per-centage which varies with the varying value of the mineral raised. It is, however, necessary to add here, that, although men who labour on this latter plan, occasionally make as much as six or ten pounds each, in a month, they are on the other hand liable to heavy losses from the speculative character of the work in which they engage. The lode may, for instance, be poor when they begin to work it, and may continue poor as they proceed farther and farther. Under these circumstances, the low value of the mineral they have raised, realizes a correspondingly low rate of per-centage; and when this happens, the best workmen cannot make more than twenty shillings a month.

Another system on which the men are employed, is the system of "contract." A certain quantity of ore in the rock is mapped out by the captain of the mine; and put up to auction among the miners thus:—One man mentions a sum for which he is willing to undertake excavating the ore, upon the understanding that he is himself to pay for the assistance, candles, &c., out of the price he asks. Another man, who is also anxious to get the contract, then offers to accept it on lower terms; a third man's demand is smaller still; and so they proceed until the piece of work is knocked down to the lowest bidder. By this sort of labour the contracting workman—after he has paid his expenses for assistance—seldom clears more than twelve shillings a week.

Upon the whole, setting his successful and his disastrous speculations fairly against each other, the Cornish miner's average gains, year by year, may be fairly estimated at about ten shillings a week. "It's hard work we have to do, sir," said my informant, summing up, when we parted, the proportions of good and evil in the social positions of his brethren and himself—" harder work than people think, down in the heat and darkness under ground. We may get a good deal at one time, but we get little enough at another; sometimes mines are shut up, and then we are thrown out altogether—but, good work or bad work, or no work at all, what with our bits of ground for potatoes and greens, and what with cheap living, somehow we and our families make it do. We contrive to keep our good cloth coat for Sundays, and go to chapel in the morning—for we're most of us Wesleyans—and then to church in the afternoon; so as to give 'em both their turn like! We never go near the mine on Sundays, except to look after the steam-pump: our rest, and our walk in the evening once a week, is a good deal to us. That's how we live, sir; whatever happens, we manage to work through, and don't complain!"

Although the occupation of smelting the copper above ground is, as may well be imagined, unhealthy enough, the labour of getting it from the mine (by blasting the subterranean rock in the first place, and then hewing and breaking the ore out of the fragments), seems to be attended with no bad effect on the constitution. The miners are a fine-looking race of men—strong and well-proportioned. The fact appears to be, that they gain more, physically, by the pure air of the cliffs and moors on which their cottages are built, and the temperance of their lives (many of them are "teetotallers"), than they lose by their hardest exertions in the under-ground atmosphere in which they work.

Serious accidents are rare in the mines of Cornwall. From the horrors of such explosions as take place in coal mines, they are by their nature entirely free. The casualties that oftenest occur are serious falls, generally produced by the carelessness of inexperienced or foolhardy people. Of these, and of extraordinary escapes from death with which they are associated, many anecdotes are told in mining districts, which would appear to the reader exaggerated, or positively untrue, if I related them on mere hearsay evidence. There was, however, one instance of a fall down the shaft of a mine, unattended with fatal consequences, which occurred while I was in Cornwall; and which I may safely adduce, for I can state some of the facts connected with the affair as an eye-witness. I attended an examination of the sufferer by a medical man, and heard the story of the accident from the parents of the patient.

On the 7th of August 1850, a boy fourteen years of age, the son of a miner, slipped into the shaft of Boscaswell Down Mine, in the neighbourhood of Penzance. He fell to the depth of thirteen fathoms, or seventy-eight feet. Fifty-eight feet down, he struck his left side against a board placed across the shaft, snapped it in two, and then falling twenty feet more, pitched on his head. He was of course taken up insensible; the doctor was sent for; and on examining him, found, to his amazement, that there was actually a chance of the boy's recovery after this tremendous fall!

Not a bone in his body was broken. He was bruised and scratched all over, and there were three cuts—none of them serious—on his head. The board stretched across the shaft, twenty feet from the bottom, had saved him from being dashed to pieces; but had inflicted at the same time, where his left side had struck it, the only injury that appeared dangerous to the medical man—a large, hard lump that could be felt under the bruised skin. The boy showed no symptoms of fever; his pulse, day after day, was found never varying from eighty-two to the minute; his appetite was voracious; and the internal functions of his body only required a little ordinary medicine to keep them properly at work. In short, nothing was to be dreaded but the chance of the formation of an abscess in his left side, between the hip and ribs. He had been under medical care exactly one week, when I accompanied the doctor on a visit to him.

The cottage where he lived with his parents, though small, was neat and comfortable. We found him lying in bed, awake. He looked languid and lethargic; but his skin was moist and cool; his face displayed no paleness, and no injury of any kind. He had just eaten a good dinner of rabbit-pie, and was anxious to be allowed to sit up in a chair, and amuse himself by looking out of the window. His left side was first examined. A great circular bruise discoloured the skin, over the whole space between the hip and ribs; but on touching it, the doctor discovered that the lump beneath had considerably decreased in size, and was much less hard than it had felt during previous visits. Next we looked at his back and arms—they were scratched and bruised all over; but nowhere seriously. Lastly, the dressings were taken off his head, and three cuts were disclosed, which even a non-medical eye could easily perceive to be of no great importance. Such were all the results of a fall of seventy-eight feet.

The boy's father reiterated to me the account of the accident, just as I had already heard it from the doctor. How it happened, he said, could only be guessed, for his son had completely forgotten all the circumstances immediately preceding the fall; neither could he communicate any of the sensations which must have attended it. Most probably, he had been sitting dangling his legs idly over the mouth of the shaft, and had so slipped in. But however the accident really happened, there the sufferer was before us—less seriously hurt than many a lad who has trodden on a piece of orange peel as he was walking along the street.

We left him (humanly speaking) certain of recovery, now that the dangerous lump in his side had begun to decrease. I heard afterwards from his medical attendant, that in two months from the date of the accident, he was at work again as usual in the mine; at that very part of it, too, where his fall had taken place!

It was not the least interesting part of my visit to the cottage where he lay ill, to observe the anxious affection displayed towards him by both his parents. His mother left her work in the kitchen to hold him in her arms, while the old dressings were being taken off and the new ones applied—sighing bitterly, poor creature, every time he winced or cried out under the pain of the operation. The father put several questions to the doctor, which were always perfectly to the point; and did the honours of his little abode to his stranger visitor, with a natural politeness and a simple cordiality of manner which showed that he really meant the welcome that he spoke. Nor was he any exception to the rest of his brother-workmen with whom I met. As a body of men, they are industrious and intelligent; sober and orderly; neither soured by hard work, nor easily depressed by harder privations. No description of personal experiences in the Cornish mines can be fairly concluded, without a collateral testimony to the merits of the Cornish miners—a testimony which I am happy to accord here; and to which my readers would cheerfully add their voices, if they ever felt inclined to test its impartiality by their own experience.

X.

THE MODERN DRAMA IN CORNWALL.

Our walk from Botallack Mine to St. Ives, led us almost invariably between moors and hills on one side, and cliffs and sea on the other; and displayed some of the dreariest views that we had yet beheld in Cornwall. About nightfall, we halted for a short time at a place which was certainly not calculated to cheer the traveller along his onward way.

Imagine three or four large, square, comfortless-looking, shut-up houses, all apparently uninhabited; add some half-dozen miserable little cottages standing near the houses, with the nasal notes of a Methodist hymn pouring disastrously through the open door of one of them; let the largest of the large buildings be called an inn, but let it make up no beds, because nobody ever stops to sleep there: place in the kitchen of this inn a sickly little girl, and a middle-aged, melancholy woman, the first staring despondently on a wasting fire, the second offering to the stranger a piece of bread, three eggs, and some sour porter corked down in an earthenware jar, as all that her larder and cellar can afford; fancy next an old, grim, dark church, with two or three lads leaning against the churchyard wall, looking out together in gloomy silence on a solitary high road; conceive a thin, slow rain falling, a cold twilight just changing into darkness, a surrounding landscape wild, barren, and shelterless—imagine all this, and you will have the picture before you which presented itself to me and my companion, when we found ourselves in the village of Morvah.

Late that night, we got to the large sea-port town of St. Ives; and stayed there two or three days to look at the pilchard fishery, which was then proceeding with all the bustle and activity denoting the commencement of a good season. Leaving St, Ives, on our way up the northern coast, we now passed through the central part of the mining districts of Cornwall. Chimneys and engine-houses chequered the surface of the landscape; the roads glittered with metallic particles; the walls at their sides were built with crystallized stones; towns showed a sudden increase in importance; villages grew large and populous; inns disappeared, and hotels arose in their stead; people became less curious to know who we were, stared at us less, gossiped with us less; gave us information, but gave us nothing more—no long stories, no invitations to stop and smoke a pipe, no hospitable offers of bed and board. All that we saw and heard tended to convince us that we had left the picturesque and the primitive, with the streets of Looe and the fishermen at the Land's End; and had got into the commercial part of the county, among sharp, prosperous, business-like people —it was like walking out of a painter's studio into a merchant's counting-house!

As we were travelling, like the renowned Doctor Syntax, in search of the picturesque, we hurried through this populous and highly-civilized region of Cornwall as rapidly as possible. I doubt much whether we should not have passed as unceremoniously through the large town of Redruth—the capital city of the mining districts—as we passed through several towns and villages before it, had not our attention been attracted and our departure delayed by a public notice, printed on rainbow-coloured paper, and pasted up in the most conspicuous part of the market-place.

The notice set forth, that "the beautiful drama of The Curate's Daughter" was to be performed at night, in the "unrivalled Sans Pareil Theatre," by "the most talented company in England," before "the most discerning audience in the world." As far as we were individually concerned, this theatrical announcement was remarkably tempting and well-timed. We were now within one day's journey of Piran Round, the famous amphitheatre where the old Cornish Miracle Plays used to be performed. Anything connected with the stage was, therefore, a subject of particular interest in our eyes. The bill before us seemed to offer a curious opportunity of studying the dramatic tastes of the modern Cornish, on the very day before we were about to speculate on the dramatic tastes of the ancient Cornish, among the remains of their public theatre. Such an occasion was too favourable to be neglected; we ordered our beds at Redruth, and joined the "discerning audience" assembled to sit in judgment on "The Curate's Daughter."

The Sans Pareil Theatre was not of that order of architecture in which outward ornament is studied.

There was nothing "florid" about it; canvas, ropes, scaffolding-poles, and old boards, threw an air of Saxon simplicity over the whole structure. Admitted within, we turned instinctively towards the stage. On each side of the proscenium boards was painted a knight in full armour, with powerful calves, weak knees, and an immense spear. Tallow candles, stuck round two hoops, threw a mysterious light on the green curtain, in front of which sat an orchestra of four musicians, playing on a trombone, an ophicleide, a clarionet, and a fiddle, as loudly as they could—the artist on the trombone, especially, performing prodigies of blowing, though he had not room enough to develop the whole length of his instrument. Every now and then great excitement was created among the expectant audience by the vehement ringing of a bell behind the scenes, and by the occasional appearance of a youth who gravely snuffed the candles all round, with a skill and composure highly creditable to him, considering the pertinacity with which he was stared at by everybody while he pursued his occupation.

At last, the bell was rung furiously for the twentieth time; the curtain drew up, and the drama of "The Curate's Daughter" began.

Our sympathies were excited at the outset. We

beheld a lady-like woman who answered to the name of "Grace;" and an old gentleman, dressed in dingy black, who personated her father, the Curate; and who was, on this occasion (I presume through unavoidable circumstances), neither more nor less than —drunk. There was no mistaking the cause of the fixed leer in the reverend gentleman's eye; of the slow swaying in his gait; of the gruff huskiness in his elocution. It appeared, from the opening dialogue, that a pending law-suit, and the absence of his daughter Fanny in London, combined to make him uneasy in his mind just at present. But he was by no means so clear on this subject as could be desired—in fact, he spoke through his nose, put in and left out his *hs* in the wrong places, and involved his dialogue in a long labyrinth of parentheses whenever he expressed himself at any length. It was not until the entrance of his daughter Fanny (just arrived from London: nobody knew why or wherefore), that he grew more emphatic and intelligible. We now observed with pleasure that he gave his children his blessing and embraced them both at once; and we were additionally gratified by hearing from his own lips, that his "daughters were the h'all on which his h'all depended—that they would watch h'over his 'ale autumn; and that whatever happened the whole party must invariably trust in heabben's obdipotent power!"

Grateful for this clerical advice, Fanny retired into the garden to gather her parent some flowers; but immediately returned shrieking. She was followed by a Highwayman with a cocked hat, mustachios, bandit's ringlets, a scarlet hunting-coat, and buff boots. This gentleman had shown his extraordinary politeness—although a perfect stranger—by giving Miss Fanny a kiss in the garden; conduct for which the Curate very properly cursed him, in the strongest language. Apparently a quiet and orderly character, the Highwayman replied by beginning a handsome apology, when he was interrupted by the abrupt entrance of another personage, who ordered him (rather late in the day, as we ventured to think) to "let go his holt, and beware how he laid his brutal touch on the form of innocence!" This newcomer, the parson informed us, was "good h'Adam Marle, the teacher of the village school." We found "h'Adam," in respect of his outward appearance, to be a very short man, dressed in a high-crowned modern hat, with a fringed vandyck collar drooping over his back and shoulders, a modern frock-coat, buttoned tight at the waist, and a pair of jack-boots of the period of James the Second. Aided by his advantages of costume, this character naturally interested us; and we regretted seeing but little of him in the first scene, from which he retired, following the penitent Highwayman out, and lecturing him as he went. No sooner were their backs turned, than a waggoner, in a clean smock-frock and high-lows, entered with an offer of a situation in London for Fanny, which the unsuspicious Curate accepted immediately. As soon as he had committed himself, it was confided to the audience that the waggoner was a depraved villain, in the employ of that notorious profligate, Colonel Chartress, who had commissioned a second myrmidon (of the female sex) to lure Fanny from virtue and the country, to vice and the metropolis. By the time the plot had "thickened" thus far, the scene changed, and we got to London at once.

We now beheld the Curate, Chartress's female accomplice, Fanny, and the vicious waggoner, all standing in a row, across the stage. The Curate, in a burst of amiability, had just lifted up his hands to bless the company, when Colonel Chartress - (dressed in an old *naval* uniform, with an opera hat of the year 1800), suddenly rushed in, followed by the Highwayman, who having relapsed from penitence to guilt, had, as a necessary consequence, determined to supplant Chartress in the favour of Miss Fanny. These two promptly seized each other by the throat; vehement shouting, scuffling, and screaming ensued; and the Curate, clasping his daughter round the waist, frantically elevated his walking-stick in the air. Was he about to inflict personal chastisement on his innocent child? Who could say? Before there was time to ask the question, the curtain fell with a bang, on the crisis of the first act.

In act the second, the first scene was described in the bills as Temple Bar by moonlight. Neither Bar nor moonlight appeared when the curtain rose—so we took both for granted, and fixed our minds on the story. The first person who now confronted us, was "good h'Adam Marle." The paint was all washed off his face; his immense spread of collar looked grievously in want of washing; and he leaned languidly on an oaken stick. He had been walking—he informed us—through the streets of London for six consecutive days and nights, without sustenance, in search of Miss Fanny, who had disappeared since the skirmish at the end of act the first, and had never been heard of since. Poor dear Marle! how eloquent he was with his white handkerchief, when he fairly opened his heart, and confided to us that he was madly attached to Fanny; that he knew he "was nothink" to her; and that, under existing circumstances, he felt inclined to rest himself on a door step! Just as he had comfortably settled down, the valet of the profligate Chartress entered, in the communicative stage of intoxication; and immediately mentioned all his master's private affairs to "h'Adam." It appeared that the Colonel had carried off Miss Fanny, had then got tired of her, and had coolly handed her over to a Jew, in part payment of "a little bill." Having ascertained the Jew's address, the indefatigable Marle left us (still without sustenance) to rescue the Curate's daughter, or die in the attempt.

The next scene disclosed Fanny, sitting conscience-stricken and inconsolable, in a red polka jacket and white muslin slip. Mr. Marle, having discovered her place of refuge, now stepped in to lecture and reclaim. Vain proceeding! The Curate's daughter looked at him with a scream, exclaimed, "Cuss me, h'Adam! cuss me!" and rushed out. "H'Adam," after a despondent soliloquy, followed with his eloquent handkerchief to his eyes; but, while he had been talking to himself, our old friend the Highwayman had been on the alert, and had picked Fanny up, fainting in the street. And what did he do with her after that? He handed her over to his "comrades in villany." And who were his comrades in villany? They were the trombone and ophicleide players from the orchestra, and the "Miss Grace," of act first, disguised as a bad character, in a cloak, with a red pocket-handkerchief over her head. And what happened next? A series of events happened next. Miss Fanny recovered on a sudden, perceived what sort of company she had about her, rushed out a second time into the street, fell fainting a second time on the pavement, and was picked up on this occasion by Colonel Chartress—in the interests, it is to be presumed, of his friend, the Jew money-lender. Before, however, he could get clear off with his prize, the indefatigably vicious Highwayman, and the indefatigably virtuous Marle, precipitated themselves on the stage, assaulting Chartress, assaulting each other, assaulting everybody. Fanny fell fainting a third time in the street; and before we could find out who was the third person who picked her up, down came the curtain in the midst of the catastrophe.

Act the third was opened by the heroine, still injured, still inconsolable, and still clad in the polka jacket and white slip. We thought her a very nice little woman, with a melodious, genteel-comedy-voice, trim ankles, and a habit of catching her breath in the most pathetic manner, at least a dozen times in the course of one soliloquy. While she was still assuring us that she felt the most forlorn creature on the face of the earth, she was suddenly interrupted by the entrance of no less a person than the Curate himself. We had seen nothing of the reverend gentleman throughout the second act; but "h'Adam" had casually informed us that his time had been passed at his parsonage, " sittun with his 'ed between his knees, sobbun!" Having now wearied of this gymnastic method of indulging in parental grief, he had set forth to seek his lost daughter, and had accidentally stopped at the very inn where she had taken refuge. Nothing could be more piteous than his present appearance; he was infinitely more tipsy, infinitely more dignified, and infinitely more parenthetical in his mode of expressing himself, than when we last beheld him. A streak of burnt cork running down each side of his venerable nose, showed us how deeply grief had increased the wrinkles of age; and our pity for him reached its climax when he cast his clerical hat on the floor, sank drowsily into a chair, and began to pray in these words: "Oh heabben! hear a solemn and a solid prayer—hear a solemn heart who wants to embrace his darling Fanny!"

All this time, the lost daughter was hiding behind the forlorn father's chair; an awful and convenient darkness being thrown on the stage by the introduction of a plank between the actors and the tallow candles. In this striking situation, Miss Fanny told her sad story, and pleaded her own cause as a stranger, under disguise of the darkness. Useless—quite useless! The reverend gentleman, having never turned round to see who it was that was speaking to him, and having therefore no idea that it was his own daughter, received in dignified silence the advances of a young person unknown to him. What course was now left to the unhappy Fanny? The old course—a rush off the stage, and a swoon in the street. As soon as her back was turned, the Parson, forgetting to take away his hat with him, staggered out at the opposite side to continue his journey. He uttered as he went the following moral observation:—"No soul so lost to Nature, but must be lost eternally—my 'art is broken!"

The next moment, we were startled by a long and elaborate trampling of feet behind the scenes, and the villain Chartress, ran panic-stricken across the stage, hotly pursued by "good h'Adam Marle." In the eloquent language of virtue, thus did Adam address him:—"Stay, ruffian, stay! Inquiring for Chartress at the bar of this inn, I found indeed that you was the very identical. You foul, venimous, treacherous, voluptuous liar, where is the un'appy Fanny? where is the victim of your prey?—Ha! 'oary-'edded ruffian, I have yer!" (*Collars Chartress.*) "But no! I will not *strike* yer; I will *drag* yer!" It was interesting to see Adam exemplify the peculiar distinction in the science of assault implied in his last words, by hauling Chartress all round the stage. It was awful to observe that the Colonel lost his temper at the second round, murderously snapped a pistol in "h'Adam's" face, and rushed off in hot homicidal triumph. We waited breathless for the fall of Marle. Nothing of the sort happened. He started, frowned, paused, laughed fiercely, exclaimed,—" The villain 'as missed !" and followed in pursuit.

In the interim, Miss Fanny had been picked up in the street, for the fourth time, by a benevolent " washerwoman," who happened to be passing by at the moment; had been conveyed to the said washerwoman's lodgings; and now appeared before us, despoiled, at last, of all the glories of the red polka, enveloped from head to foot in clouds of white muslin, and dying with frightful rapidity in an armchair. In the next and last scene, all that remained to represent the unhappy heroine was a coffin decently covered with a white sheet. With slow and funereal steps, the Curate, Miss Grace, "h'Adam," the Highwayman, and the "venimous and voluptuous liar," Chartress, approached to weep over it. The Curate had gone raving mad since we saw him last. His wig was set on wrong side foremost; the ends of his clerical cravat floated wildly, a yard long at least over his shoulders; his eyes rolled in frenzy; he swooned at the sight of the coffin; recovered convulsively; placed Marle's hand in the hand of Miss Grace (telling him that now one daughter was dead, nothing was left for him but to marry the other); and then fell flat on his back, with a thump that shook the stage and made the audience start unanimously. Marle—well-bred to the last—politely offered his arm to Grace; and pointing to the coffin, asked Chartress, reproachfully, whether that was not *his* work. The Colonel took off his opera-hat, raised his hand to his eyes, and doggedly answered, "Indeed, it is!" The Tableau thus formed, was completed by the Highwayman, the coffin, and the defunct Curate; and the curtain fell to slow music.

Such was the plot of this remarkable dramatic work, exactly as I took it down in the theatre, between the acts; noting also in my pocket-book such scraps of dialogue as I have presented to the reader, while they fell from the actors' lips. There were plenty of comic scenes in the play which I leave unmentioned; for their humour was of the dreariest, and their morality of the lowest order that can possibly be conceived. I can only say, as the result of my own experience at Redruth, that if the dramatic reforms which are now being attempted in the theatrical by-ways of the metropolis succeed, there would be no harm in extending the experiment as far as the locomotive stage of Cornwall. Good plays are good missionaries; and, like missionaries, let them travel to teach.

And now, having seen enough of the modern drama in Cornwall, without waiting for the songs, the dances, and the farces which are to follow the "Curate's Daughter," let us go on to Piranzabuloe, and look at the theatre in which the Cornish of former days assembled; endeavouring to discover, at the same time, by what sort of performances the people were instructed or amused some two hundred and fifty years ago.

XI.

THE ANCIENT DRAMA IN CORNWALL.

WE found the modern Cornish theatre situated in a populous town; built up, as a temporary structure, with old canvas and boards; and opened to audiences only at night. We found the ancient Cornish theatre placed in a perfect desert; constructed permanently, though rudely, of mounds of turf—the sky forming its only roof, the flat plain its only stage, the broad daylight its only means of illumination. Nothing of the kind could be more strongly marked than the difference between the theatre of the past, and the theatre of the present day, in the far West of England.

In like manner, the country about Piran Round (such is the name of the Old Cornish amphitheatre) offers a startling contrast to the country about Redruth. You are at once powerfully impressed by its barren solitude, its dreary repose, after the fertility and populousness of the great mining districts through which you have just passed. Now, the large towns and busy villages disappear, the mines grow rarer, the roads look deserted, the wide pathways dwindle to the merest foot-track. Again you behold the spacious moor rolling away in alternate hill and dale to the far horizon; again you pass though the quaint coast villages, and see the few simple cottages, the few old boats, the little groups talking quietly at the inn door, as they have already presented themselves along the southern and western shores of Cornwall. Soon, however, your onward road towards Piran Round becomes yet more desolate. Ere long, not even a solitary cottage is in sight, not a living being appears: you find yourself wandering along the uneven boundary of a wilderness of sand-hills heaped up from the seashore by the wind. You look over a perfect desert of miniature mountains and valleys, in some places overgrown with thin, dry grass; in others, dotted with little pools of mud and stagnant water. Year by year, this invasion of sand encroaches on the moorland—year by year, it is ever shifting, ever increasing, ever assuming newer and more fantastic forms, now in one direction and now in another, with each fresh storm.

When you leave this dreary scene, you only leave it for the wild flat heath, the open naked country once more. You follow your long road, visible miles on before you, winding white and serpent-like over the dark ground, until you suddenly observe in the distance an object which rises strangely above the level prospect. You approach nearer, and behold a circular turf embankment; a wide, lonesome, desolate enclosure, looking like a witches' dancing-ring that has sprung up in the midst of the open moor. This is Piran Round. Here, the old inhabitants of Cornwall assembled to form the audience of the drama of former days.

A level area of grassy ground, one hundred and thirty feet in diameter, is enclosed by the embankment. There are two entrances to this area cut through the boundary circle of turf and earth, which rises to a height of nine or ten feet, and narrows towards the top, where it is seven feet wide. All round the inside of the embankment steps were formerly cut; but their traces are now almost obliterated by the growth of the grass. They were originaly seven in number; the spectators stood on them in rows, one above another—a closely packed multitude, all looking down at the dramatic performances taking place on the wide circumference of the plain. When it was well filled, the amphitheatre must have contained upwards of two thousand people.

Such is this rude, yet extraordinary structure, in our time. It has not lost its patriarchal simplicity since the far distant period when the populace thronged its turf steps to welcome the strolling players of their age. The antiquity of Piran Round dates back beyond the period of the earliest and rudest dramatic performances on English ground. It was first used for popular sports, for single combats, for rustic councils. Then, plays were acted in it— miracle plays — some translated into the ancient Cornish language, some originally written in it. The oldest of these are lost; but one of a comparatively late date has been preserved and translated into English. We will examine this book while we sit within the deserted amphitheatre; and thus, in imagination at least, people the simple stage before us with the rough country actors who once trod it—thus pry behind the scenes at all that is left to us of the ancient drama in Cornwall.

The play which we now open is called by the comprehensive title of "The Creation of the World, with Noah's Flood." It was translated in 1611, from a drama of much earlier date, for performance in Cornish, by William Jordan; was then rendered into English by John Keygwyn, in 1691; and was finally corrected and published by Mr. Davies Gilbert, in 1827. The Cornish and English versions are printed on opposite pages, so we can compare the two throughout, as we go on.

The play is in five acts, and is written in poetry —in a rambling octosyllabic metre, often varied by the introduction of longer or shorter lines, and sometimes interspersed (in the Cornish version) with a word or two of English. It occupies a hundred and eighty pages, containing on the average about twenty-five lines each. This would be thought rather a lengthy manner of developing a dramatic story in our days; but we must remember that the time embraced in the plot of the old playwright extends from the Creation to the Flood, and must be astonished and thankful that he has not been more diffuse.

The *dramatis personæ* muster by the legion. In the first act, we have the whole heavenly host: in the second, are superadded Adam, Eve, "Torpen, a devil," Beelzebub, the Serpent, and Michael the Archangel; in the third, besides these, Death, Cain and his wife, Abel and Seth; in the fourth, we have the addition of Lamech, a servant, a Cherubim, and a first and second devil; and in the fifth, Enoch, Noah and his wife, Shem, Ham, Japhet, Seth, Jaball, and Tubal Cain.

The author manages this tremendous list of mortal and immortal characters with infinite coolness and dexterity. Nothing appears to embarrass him. He follows history in a negligent, sauntering way, passing over a hundred years or so, whenever it is convenient; and giving all his personages their turn of talking in orderly and impartial rotation. His speeches are wonderfully moral and long; even his worst characters have, for the most part, a temperate and logical way of uttering the most violent language, which must have read an excellent lesson to the roistering young gentlemen among the audiences of the time.

We will now examine the play a little in detail, quoting the stage directions (the most extraordinary part of it) exactly as they occur; and occasionally presenting a line or two of the dialogue from the old

English translation wherever it best illustrates the author's style.

The first act comprehends the fall of the angels—the introductory stage direction commanding that the theatrical clouds, and the whole sky to boot, shall open when Heaven is named! All is harmony at the outset of the play, until it is Lucifer's turn to speak. He declares that he alone is great, and that all allegiance must be given to him. Some of the angels glorify him accordingly; others remain true to their celestial service; the debate grows warm, and some of the disputants give each other the lie (but very calmly). At length, the scene is closed by Lucifer's condemnation to Hell, which, as the directions provide, "shall gape when it is named." The faithful angels are then told to "have swords and staves ready for Lucifer," who, we are informed, "voideth and goeth down to Hell apparelled foul, with fire about him, turning to Hell, with every degree of devils and lost spirits on cords running into the plain." With this stirring scene the act ends.

The second act comprises the creation and fall of man. Here, again, we will consult the stage directions, as giving the best idea of the incidents and scenes. We find that Adam and Eve are to be "apparelled in white leather in a place appointed by the conveyor" (probably the person we term stage-manager now); "and are not to be seen until they be called; and then each rises." After this, we read:—"Let Paradise be finely made, with fair trees in it, and apples upon a tree, and other fruit on the others. A fountain, too, in Paradise, and fine flowers painted. Put Adam into Paradise—let flowers appear in Paradise—let Adam lie down and sleep where Eve is, and she, by the conveyor, must be taken from Adam's side—let fishes of all sorts, birds and beasts, as oxen, kyne, sheep, and such like, appear."

Then, we have the preparations for the temptation, ordered thus:—"A fine serpent to be made with a virgin's face, and yellow hair on her head. Let the serpent appear, and also geese and hens." Lucifer enters immediately afterwards, and goes into the serpent, which is then directed to be "seen singing in a tree" (the actor who personated Lucifer must have had some gymnastic difficulties to contend with in his part!)—" Eve looketh strange on the serpent;" then, "talketh familiarly and cometh near him;" then, "doubteth and looketh angrily;" and then eats part of the apple, shows it to Adam, and insists on his eating part of it too, in the following lines :—

> "Sir, in a few words,
> Taste thou part of the apple,
> Or my love thou shalt lose!
> See, take this fair apple,
> Or surely between thee and thy wife
> The love shall utterly fail,
> If thou wilt not eat of it!"*

The stage direction now proceeds :—"Adam receiveth the apple and tasteth it, and so repenteth and casteth it away. Eve looketh on Adam very strangely

---

* In case any of my readers should feel desirous of seeing a specimen of the Cornish language at the date of the play, I subjoin the original text of the seven lines of John Keygwyn's translation, quoted above.

> "Syr, war nebas lavarow,
> Tast gy part an avallow,
> Po ow harenga ty a gyll!
> Meir, Kymar an avall teake,
> Po sure inter te ha'th wreage
> An garenga quyt a fyll
> Mar ny vynyth y thebbry!"

Some of this looks like a very polyglot language. But the ancient Cornish tongue had altered and deteriorated; and was indeed changing into English at the period of our play. Why the author should have helped himself, in his literary emergency, to the two Latin words in the fifth line (*inter te*) and speaketh not anything." During this pause, the "conveyor" is told " to get the fig-leaves ready." Then Lucifer is ordered to "come out of the serpent and creep on his belly to hell ;" Adam and Eve receive the curse, and depart out of Paradise, "showing a spindle and distaff"—no badly-conceived emblem of the labour to which they are henceforth doomed. And thus the second act terminates.

The third act treats of Cain and Abel; and is properly opened by an impersonation of Death. After which Cain and Abel appear to sacrifice.

Cain makes his offering of the first substance that comes to hand—"dry cow-dung" (!) ; and tells Abel that he is a "dolthead" and " a frothy fool" for using anything better. " Abel is stricken with a jawbone and dieth; Cain casteth him into a ditch." The effect of the first murder on the minds of our first parents, is delineated in some speeches exhibiting a certain antique simplicity of thought, which almost rises to the poetical by its homely adherence to nature, and its perfect innocence of effort, artifice, or display. The

---

when English would have served his turn as well, it is difficult to discover, unless he wished to show his learning before the rustic audiences of Piran Round.

banishment of Cain, still glorying in his crime, follows the lamentations of Adam and Eve for the death of Abel; and the act is closed by Adam's announcement of the birth of Seth.

The fourth act relates the deaths of Cain and Adam, and contains some of the most eccentric, and also, some of the most elevated writing in the play. Lamech opens the scene, candidly and methodically exposing his own character in these lines:—

> "Sure I am the first
> That ever yet had two wives!
> And maidens in sufficient plenty
> They are to me. I am not dainty,
> I can find them where I will;
> Nor do I spare of them
> In anywise one that is handsome.
> But I am wondrous troubled,
> Scarce do I see one glimpse
> What the devil shall be done!"

In this vagabond frame of mind Lamech goes out hunting, with bow and arrow, and shoots Cain, accidentally, in a bush. When Cain falls, Lamech appeals to his servant, to know what is it that he has shot. The servant declares that is it "hairy, rough, ugly, and a buck-goat of the night." Cain, however, discovers himself before he dies. There is something rudely dreary and graphic about his description of his loneliness, bare as it is of any recommendation of metaphors or epithets:

> "Deformed I am very much,
> And overgrown with hair;
> I do live continually in heat or cold frost,
> Surely night and day;
> Nor do I desire to see the son of man,
> With my will at any time;
> But accompany most time with all the beasts."

Lamech, discovering the fatal error that he has committed, kills his servant in his anger; and the scene ends with "the devils carrying them away with great noise to hell."

The second scene is between Adam and his son Seth; and here, the old dramatist often rises to an elevation of poetical feeling, which, judging from the preceding portions of the play, we should not have imagined he could reach. Barbarous as his execution may be, the simple beauty of his conception often shines through it faintly, but yet palpably, in this part of the drama.

Adam is weary of life and weary of the world; he sends Seth to the gates of Paradise to ask mercy and release for him, telling his son that he will find the way thither by his father's foot-prints, burnt into the surface of the earth which was cursed for Adam's transgression. Seth finds and follows the supernatural marks, is welcomed by the angel at the gate of Paradise, and is permitted to look in. He beholds there, an Apocalypse of the redemption of the world. On the tree of life sit the Virgin and Child; while on the tree from which Eve plucked the apple, "the woman" is seen, having power over the serpent. The vision changes, and Cain is shown in hell, "sorrowing and weeping." Then the angel plucks three kernels from the tree of life, and gives them to Seth for his father's use, saying that they shall grow to another tree of life, when more than five thousand years are ended; and that Adam shall be redeemed from his pains when that period is fulfilled. After this, Seth is dismissed by the angel and returns to communicate to his father the message of consolation which he has received.

Adam hears the result of his son's mission with thankfulness; blesses Seth; and speaks these last words, while he is confronted by Death:—

> "Old and weak, I am gone!
> To live longer is not for me:
> Death is come,
> Nor will here leave me
> To live one breath!

rudely dreary and graphic about his description of his loneliness, bare as it is of any recommendation of metaphors or epithets:

> "Deformed I am very much,
> And overgrown with hair;
> I do live continually in heat or cold frost,
>   Surely night and day;
> Nor do I desire to see the son of man,
> With my will at any time;
> But accompany most time with all the beasts."

Lamech, discovering the fatal error that he has committed, kills his servant in his anger; and the scene ends with "the devils carrying them away with great noise to hell."

The second scene is between Adam and his son Seth; and here, the old dramatist often rises to an elevation of poetical feeling, which, judging from the preceding portions of the play, we should not have imagined he could reach. Barbarous as his execution may be, the simple beauty of his conception often shines through it faintly, but yet palpably, in this part of the drama.

Adam is weary of life and weary of the world; he sends Seth to the gates of Paradise to ask mercy and release for him, telling his son that he will find the way thither by his father's foot-prints, burnt into the surface of the earth which was cursed for Adam's transgression. Seth finds and follows the supernatural marks, is welcomed by the angel at the gate of Paradise, and is permitted to look in. He beholds there, an Apocalypse of the redemption of the world. On the tree of life sit the Virgin and Child; while on the tree from which Eve plucked the apple, "the woman" is seen, having power over the serpent. The vision changes, and Cain is shown in hell, "sorrowing and weeping." Then the angel plucks three kernels from the tree of life, and gives them to Seth for his father's use, saying that they shall grow to another tree of life, when more than five thousand years are ended; and that Adam shall be redeemed from his pains when that period is fulfilled. After this, Seth is dismissed by the angel and returns to communicate to his father the message of consolation which he has received.

Adam hears the result of his son's mission with thankfulness; blesses Seth; and speaks these last words, while he is confronted by Death:—

> "Old and weak, I am gone!
> To live longer is not for me:
> Death is come,
> Nor will here leave me
> To live one breath!

> I see him now with his spear,
> Ready to pierce me on every side,
> There is no escaping from him!
> The time is welcome with me—
> I have served long in the world!"

So, the patriarch dies, trusting in the promise conveyed through his son; and is buried by Seth "in a fair tomb, with some Church sonnet."

After this impressive close to the fourth act—impressive in its intention, however clumsy the appliances by which that intention is worked out—it would be doing the old author no kindness to examine his fifth act in detail. Here, he sinks again in many places, to puerility of conception and coarseness of dialogue. It is enough to say that the history of the Flood closes the drama, and that the spectators are dismissed with an epilogue, directing them to "come to-morrow, betimes, and see very great matters"—the minstrels being charged, at the conclusion to "pipe," so that all may dance together, as the proper manner of ending the day's amusements.

And now, let us close the book, look forth over this lonesome country and lonesome amphitheatre, and imagine what a scene both must have presented, when a play was to be acted on a fine summer's morning in the year 1611.

Fancy, at the outset, the arrival of the audience—people dressed in the picturesque holiday costume of the time, which varied with every varying rank, hurrying to their daylight play from miles off; all visible in every direction on the surface of the open moor, and all converging from every point of the compass to the one common centre of Piran Round. Then, imagine the assembling in the amphitheatre; the running round the outer circle of the embankment to get at the entrances; the tumbling and rushing up the steps inside; the racing of hot-headed youngsters to get to the top places; the sly deliberation of the elders in selecting the lower and safer positions; the quarrelling when a tall man chanced to stand before a short one; the giggling and blushing of buxom peasant wenches when the gallant young bachelors of the district happened to be placed behind them; the universal speculations on the weather; the universal shouting for pots of ale—and finally, as the time of the performance drew near and the minstrels appeared with their pipes, the gradual hush and stillness among the multitude; the combined stare of the whole circular mass of spectators on one point in the plain of the amphitheatre, where all knew that the actors lay hidden in a pit, properly covered in from observation

— the mysterious "green-room" of the strolling players of old Cornwall!

And the play!—to see the play must have been a sight indeed! Conceive the commencement of it; the theatrical sky which was to open awfully whenever Heaven was named; the mock clouds coolly set up by the "property-man" on an open-air stage, where the genuine clouds appeared above them to expose the counterfeit; the hard fighting of the angels with swords and staves; the descent of the lost spirits along cords running into the plain; the thump with which they must have come down; the rolling off of the whole troop over the grass, to the infernal regions, amid shouts of applause from the audience as they rolled! Then the appearance of Adam and Eve, packed in white leather, like our modern dolls—the serpent with the virgin's face and the yellow hair, climbing into a tree, and singing in the branches— Cain falling out of the bush when he was struck by the arrow of Lamech, and his blood appearing, according to the stage directions, when he fell—the making of the Ark, the filling it with live stock, the scenery of the Deluge, in the fifth act! What a combination of theatrical prodigies the whole performance must have presented! How the actors must have ranted to make themselves heard in the open air; how often the machinery must have gone wrong, and the rude scenery toppled and tumbled down! Could we revive at will, for mere amusement, any of the bygone performances of the theatre, since the first days of barbaric acting in a cart, assuredly the performances at Piran Round would be those which, without hesitation, we should select from all others to call back to life.

The end of the play, too—how picturesque, how striking all the circumstances attending it must have been! Oh that we could hear again the merry old English tune piped by the minstrels, and see the merry old English dancing of the audience to the music! Then, think of the separation and the return home of the populace, at sunset; the fishing people strolling off towards the seashore; the miners walking away farther inland; the agricultural labourers spreading in all directions, wherever cottages and farm-houses were visible in the far distance over the moor. And then the darkness coming on, and the moon rising over the amphitheatre, so silent and empty, save at one corner, where the poor worn-out actors are bivouacking gipsy-like in their tents, cooking supper over the fire that flames up red in th moonlight, and talking languidly over the fatigues and the triumphs of the play. What a moral and what a beauty in the quiet night-view of the old amphitheatre, after the sight that it must have presented during the noise, the bustle, and the magnificence of the day!

Shall we dream over our old play any longer? Shall we delay a moment more, ere we proceed on our journey, to compare the modern with the ancient drama in Cornwall, as we have already compared the theatre of Redruth with the theatre of Piran Round? If we set them fairly against one another as we now know them, would it be rash to determine which burnt purest—the new light that flared brilliantly in our eyes when we last saw it, or the old light that just flickered in the socket for an instant, as we tried to trim it afresh? Or, if we rather inquire which audience had the advantage of witnessing the worthiest performance, should we hesitate to decide at once? Between the people at Redruth, and the people at Piran Round, there was certainly a curious resemblance in one respect—they failed alike to discern the barbarisms and absurdities of the plays represented before them; but were they also equally uninstructed by what they beheld?

Which was likeliest to send them away with something worth thinking of, and worth remembering—the drama about knaves and fools, at the modern theatre, or the drama about Scripture History at the ancient? Let the reader consider and determine.

For our parts, let us honestly confess that though we took up the old play (not unnaturally) to laugh over the clumsiness and eccentricity of the performance, we now lay it down (not inconsistently), recognising the artless sincerity and elevation of the design—just as in the earliest productions of the Italian School of Painting we first perceive the false perspective of a scene or the quaint rigidity of a figure, and only afterwards discover that these crudities and formalities enshrine the germs of deep poetic feeling, and the first struggling perceptions of grace, beauty, and truth.

XII.

THE NUNS OF MAWGAN.

About three miles from the large market-town of St. Colomb Major, in the direction of the coast, is situated the Vale of Mawgan. The village of the same name occupies the lower part of the valley, and includes a few cottages, an old church, a yet older manor-house, and a clear running stream, crossed by a little stone bridge, all nestling close together on a few hundred yards of ground enclosed by some of the most luxuriant wood foliage in Cornwall. The trees bound each side of the stream, tinging it in deep places where it eddies smoothly, with hues of lustrous green; and dipping their lower branches into it, where it ripples on white pebbles or glides fast over grey sand. They cluster thickly about the old church-yard, as if to keep the place secret, throwing deep shadows over the graves, and hiding all outer objects from the eye. The small cottage garden and the spacious manor-house enjoy their verdant shelter alike; the bye-roads leading in and out of the village, are soon lost to view amid outspread branches; and not even a peep of the land that leads on to the sea in one direction, and back to the town in the other, is to be obtained through the natural screen of leaves above, and mosses, ferns, and high grass below, which closely shut in this part of the Vale of Mawgan from the open country around.

There is an unbroken, unworldly tranquillity about this secluded place, which communicates itself mysteriously to the stranger's thoughts; making him unconsciously slacken in his walk, and look and listen in silence, when he enters it, as if he had penetrated into a new sphere. Slight noises, rarely noticed elsewhere, are always audible here. The dull fall of the latch, when an idle child carelessly opens the churchyard wicket, sounds from one end of the village to the other. The curious traveller who wanders round the walls of the old church, peering through its dusty lattice windows at the dark religious solitude within, can hear the lightest flap of a duck's wing in the stream below; or the gentlest rustle of distant leaves, as the faint breeze moves them in the upland woods above. But these, and all other sounds, never break the peaceful charm of the place—they only deepen its unearthly stillness.

Within the church-yard, the bright colour of the turf, and the quiet grey hues of the mouldering tombstones, are picturesquely intermingled all over the uneven surface of the ground, save in one remote corner, where the graves are few and the grass grows rank and high. Here, the eye is abruptly attracted by the stern of a boat, painted white, and fixed upright in the earth. This strange memorial, little suited though it be to the old monuments around, has a significance of its own which gives it a peculiar claim to consideration. Inscribed on it, appear the names of ten fishermen of the parish who went out to sea to pursue their calling, on one wintry night in 1846. It was unusually cold on land—on the sea, the frosty bitter wind cut through to the bones. The men were badly provided against the weather; and hardy as they were, the weather killed them that night. In the morning, the boat drifted on shore, manned like a spectre bark, by the ghastly figures of the dead—freighted horribly with the corpses of ten men all frozen to death. They are now buried in Mawgan church-yard; and the stern of the boat they died in tells their fatal story, and points to the last home which they share together.

But it is not from such a village tragedy as this; it is not from its retired situation, its Arcadian peacefulness, its embowering trees and hidden hermit-like beauties of natural scenery, that the vale of Mawgan derives its peculiar interest. It possesses an additional attraction, stronger than any of these, to fix our attention—it is the scene of a romance which we may still study, of a mystery which is of our own time. Even to this little hidden nook, even to this quiet bower of Nature's building, that vigilant and indestructible Papal religion, which defies alike hidden conspiracy and open persecution, has stretched its stealthy and far-spreading influence. Even in this remote corner of the remote west of England, among the homely cottages of a few Cornish peasants, the imperial Christianity of Rome has set up its sanctuary in triumph—a sanctuary not thrown open to dazzle and awe the beholder, but veiled in deep mystery behind gates that only open, like the fatal gates of the grave, to receive, but never to dismiss again to the world without.

It is this attribute of the vale of Mawgan which leads the stranger away from the cool, clear stream, and the pleasant, shadowy recesses among the trees, to an ancient building near the church, which he knows to have been once an old English manorial hall—to be now a convent of Carmelite nuns.

The House of Lanhearne, so it is named, comprises an ancient and a modern portion; the first dating back before the time of the Conquest, the second added probably not more than a century and a half ago. The place formerly belonged to the old Cornish family of the Arundels; but about the year 1700, their race became extinct, and the property passed into the possession of the present Lord Arundel. However, although the manor-house has changed masters, there is one peculiar circumstance connected with it, which has remained unaltered down to the present time—it has never had a Protestant owner.

Thus, whatever religious traditions are connected with it, are Roman Catholic traditions. A secret recess remains in the wall of the old house, where a priest was hidden from his pursuers, during the reign of Elizabeth, for eighteen months; the place being only large enough to allow a man to stand upright in it. The skull of another priest who was burnt at the same period, is also preserved with jealous care, as one of the important relics of the ancient history of Lanhearne.

About the commencement of this century, the manor-house entirely changed its character. It was at that time given to the Carmelite nuns, who now inhabit it, by Lord Arundel. The sisterhood was originally settled in France, and was removed thence to Antwerp, at the outbreak of the first French Revolution. Shortly afterwards, when the affairs of the Continent began to assume a threatening and troubled aspect, the nuns again migrated, and sought in England, at Lanhearne House, the last asylum which they still occupy.

The strictness of their order is preserved with a severity of discipline which is probably without parallel anywhere else in Europe. It is on our free English ground, in one of our simplest and prettiest English villages, that the austerities of a Carmelite convent are now most resolutely practised, and the seclusion of a Carmelite convent most vigilantly preserved, by the nuns of Mawgan! They are at present twenty in number: two of them are Frenchwomen, the rest are all English. They are of every age, from the very young to the very old. The eldest of the sisterhood has long passed the ordinary limits of human life—she has attained ninety-five years.

The nuns never leave the convent, and no one even sees them in it. Women even are not admitted to visit them: the domestic servants, who have been employed in the house for years, have never seen their faces, have never heard them speak. It is only in cases of severe and dangerous illness, when their own skill and their own medicines do not avail them, that they admit, from sheer necessity, the only stranger who ever approaches them—the doctor; and on these occasions, whenever it is possible, the face of the patient is concealed from the medical man.

The nuns occupy the modern part of the house, which is entirely built off, inside, from the ancient. Their only place for exercise is a garden of two acres, enclosed by lofty walls, and surrounded by trees. Their food and other necessaries are conveyed to them through a turning door; all personal communition with the servants' offices being carried on through the medium of lay sisters. The nuns have a private way, known only to themselves, to the chapel choir, which is constructed in the form of a gallery, boarded in at the sides and concealed by a curtain and close grating in front. The chapel itself is in the old part of the house, and occupies what was formerly the servants' hall. The officiating priest who undertakes the duties here, lives in this portion of the building, and leads a life of complete solitude, until he is relieved by a successor. He never sees the face of one of the nuns; he cannot even ask one of his own profession to dine with him, without first of all obtaining (by letter) the express permission of the Abbess; and when his visitor is at length admitted, it is impossible to gain for him—let him be who he may—the additional indulgence of being allowed to sleep in the house.*

The chapel is the only part of the whole interior of the building to which strangers can be admitted: those who desire to do so can attend mass there on Sundays. The casual visitor, when permitted to enter it, is not allowed to pass beyond the pillars which support the gallery of the choir above him;

---

* All the particulars here related of the convent discipline, were communicated to me by the resident priest. This gentleman was certainly not a prejudiced witness on the side of austerity—for he frankly complained of the lonely life which the rules of the Sisterhood inflicted on him, and unhesitatingly acknowledged that he was anxious for the time when his clerical successor would come to relieve him.

for if he advanced farther, the nuns who might then be occupying it, might see him while they were engaged at their devotions. The chapel exhibits nothing in the way of ornament, beyond the altar furniture and a few copies from pictures on sacred subjects by the old masters. Some of the more valuable objects devoted to its service are not shown. These consist of the sacred vestments and the sacramental plate, which are said to be of extraordinary beauty and value, and are preserved in the keeping of the Abbess. The worth of one of the jewelled chalices alone has been estimated at a thousand pounds.

Much of the land in the neighbourhood belongs to the convent, which has been enriched by many valuable gifts. The nuns make a good use of their wealth. Neither the austerities and mortifications to which their lives are devoted, nor their rigid and terrible self-exclusion from all intercourse with their fellow-beings in the world around them, have diminished their sympathy for affliction, or their readiness in ministering to the wants of the poor. Any assistance of any kind that they can render, is always at the service of those who require it, without distinction of rank or religion. No wandering beggar who rings at the convent bell, ever leaves the door without a penny and a piece of bread to help him on his way.

But the charities of the nuns of Mawgan do not stop short at the first good work of succouring the afflicted; they extend also to a generous sympathy for those human weaknesses of impatience and irresolution in others, which they have surmounted, but not forgotten themselves. Rather more than twelve years since, a young girl of eighteen applied to be admitted to share the dreary life-in-death existence of the Carmelite sisterhood. She was received for her year of probation: it expired, and she still held firmly to her first determination. But the nuns, in pity to her youth, and perhaps mournfully remembering, even in their life-long seclusion of mind and body, how strong are the ties which bind together the beings of this world and the things of this world, gave her more time yet to search her own motives, to look back on what she was abandoning, to look forward on what she desired to obtain. Mercifully refusing to grant her her own wishes, they forebore the performance of the fatal ceremony which irrevocably took her from earth to give her up only to Heaven, until she had undergone an additional year of probation. This last solemn period of delay which Christian charity and sisterly love had piously granted, expired, and found her still determined to adhere to her resolution. She took the veil; and the dreary gates of Lanhearne have closed on all that is mortal of her for ever!

The convent has two burial places. The first is in an ancient recess within the village church, and was given to the nuns with the manor-house. Those among them who first expired on English ground, lie buried here—the Catholic dead have returned to the once Catholic edifice, where the Protestant living now worship! When the Carmelite funeral procession entered this place, it entered at the dead of night, to avoid the chance of any intrusion. But as the nuns have no private entrance to their burial-vault, and have been by law prohibited from making one; as they are obliged to pass through the public door of the church and walk up the nave, they are at the mercy of any stranger who can gain admittance to the building, and who may be led by idle curiosity to watch the ceremonies which accompany their midnight service for the dead. Feeling this, they have of late years abandoned their burial place, after first carefully boarding it off from all observation. No inquisitive eyes can now behold, no intruding footsteps can now approach, the tombs of the nuns of Mawgan.

The second cemetery, which they use at present, is situated in one of the convent-gardens, and can therefore be secured, whenever they please, from all observation. A wooden door at one corner of the ancient portion of the manor-house leads into it. The place is merely a small, square plot of ground, damp, shady, and overgrown with long grass. An old and elaborately carved stone cross stands in it; and about this appear the graves of the nuns, marked by plain slate tablets. But even here, the mystery which hangs darkly over the Carmelite household does not clear—the seclusion that has hidden the living in the Convent, is but the forerunner of the secrecy that veils from us on the tombstone the history of the dead. The saint's name once assumed by the nun, and the short yet beautiful supplication of the Roman Church for the repose of the soul of the departed, form the only inscriptions that appear over the graves.

This is all—all of the lives, all of the deaths of the

sisterhood at Lanhearne that we can ever know! The remainder must be conjecture. We have but the bare stern outline that has been already drawn— who shall venture, even in imagination, to colour and complete the picture which it darkly, yet plainly, indicates?

Even if we only endeavour to image to ourselves the externals of the life which those massy walls keep secret, what have we to speculate on? Nothing but the day that in winter and summer, in sunshine and in storm, brings with it year after year, to young and to old alike, the same monotony of action and the same monotony of repose—the turning door in the wall (sole indication to those within, that there is a world without), moved in silence, ever at the same stated hour, by invisible hands—the prayer and penance in the chapel choir, always a solitude to its occupants, however many of their fellow-creatures may be standing beneath it—the short hours of exercise amid high garden walls, which shut out everything but the distant sky. Beyond this, what remains but that utter vacancy where even thought ends; that utter gloom in which the brightest fancy must cease to shine?

Should we try to look deeper than the surface;

to strip the inner life of the convent of all its mysteries and coverings, and anatomising it inch by inch, search it through down to the very heart? Should we pry into the dread and secret processes by which, among these women, one human emotion after another may be suffering, first ossification, then death? No!—this is a task which is beyond our power; an investigation which, of our own knowledge, we cannot be certain of pursuing aright. We may imagine grief that does not exist, remorse that is not felt, error that has not been committed. It is not for us to criticise the catastrophe of the drama, when we have no acquaintance with the scenes which have preceded it. It is not for us, guided by our own thoughts, moved by the impulses of the world we live in, to decide upon the measure of good or evil contained in an act of self-sacrifice at the altar of religion, which is in its own motive and result so utterly separated from all other motives and results, that we cannot at the outset even so much as sympathise with it. The purpose of the convent system is of those purposes which are conceived in this world, but which appeal for justification or condemnation only to the next.

"Judge not, that ye be not judged!" Those words sink deep into our hearts, as we look our last upon the convent walls, and leave the living-dead at old Lanhearne.

XIII.

LEGENDS OF THE NORTHERN COAST.

From the time when we left St. Ives, we walked through the last part of our journey much faster than we walked through the first; faster, perhaps, than the reader may have perceived from these pages. When we stopped at the town of St. Columb Major, to visit the neighbouring vale of Mawgan, we had already advanced half way up the northern coast of Cornwall. Throughout this part of the county the towns lay wide asunder; and, as pedestrian tourists, we were obliged to lengthen our walks and hasten our pace accordingly.

After we had quitted St. Columb Major, our rambles began to draw rapidly to their close. Little more was now left for us to examine than the different localities connected with certain interesting Cornish legends. The places thus associated with the quaint fancies of the olden time, were all situated close together, some fifteen or twenty miles farther on, along the coast. The first among them that we reached was Tintagel Castle, an ancient ruin magnificently situated on a precipice overhanging the sea, and romantically, if not historically, reputed as the birthplace of King Arthur.

The date of the Castle of Tintagel is as much a subject of perplexity among modern antiquaries, as is the existence of King Arthur among modern historians. We may still see some ruins of the Castle; but when or by whom the building was erected which those ruins represent, we have no means of discovering: we only know that, after the Conquest, it was inhabited by some of our English princes, and that it was used as a state prison so late as the reign of Elizabeth. The rest is, for the most part, mere conjecture, raised upon the weak foundation of a few mouldering fragments of walls which must soon crumble and disappear as the rest of the Castle has crumbled and disappeared before them.

The position of the old fortress was, probably, almost impregnable in the days of its strength and glory. The outer part of it was built on a precipitous projection of cliff, three hundred feet high, which must have been wrenched away from the mainland by some tremendous convulsion of Nature. The inner part stood on the opposite side of the chasm formed by this convulsion; and both divisions of the fortress were formerly connected by a draw-bridge. The most interesting portion of the few ruins now remaining, is that on the outermost promontory, which is almost entirely surrounded by the sea. The way up to this cliff is by a steep and somewhat perilous path; so narrow in certain places, where it winds along the verge of the precipice, that a single false step would be certain destruction. The difficulties of the ascent appear to have impressed the old historian of Cornwall, Norden, so vividly that he tries in his "Survey," to frighten all his readers from attempting it; warning "unstable man," if he will try to mount the cliff, that "while he respecteth his footinge he indaungers his head; and looking to save the head, indaungers the footinge, accordinge to the old proverbe: *Incidit in Scyllam qui vult vitare Charybdim.* He must have eyes,"—ominously adds the worthy Norden—"that will scale Tintagell."

The ruins on the summit of the promontory only consist of a few straggling walls, loosely piled up, healed of all his wounds in the Fairy Land; and that he would yet return to lead and to govern them, as of old.

Such is the scene—strange compound of fiction and truth, of the typical and the real—which legends teach us to imagine in the Tintagel Castle of thirteen centuries ago! What is the scene that we look on now?—A solitude where the decaying works of man, and the enduring works of Nature appear mingled in beauty together. The grass grows high and luxuriant, where the rushes were strewn over the floor of Arthur's banqueting hall. Sheep are cropping the fresh pasture, within the walls which once echoed to the sweetest songs, or rang to the clash of the stoutest swords of ancient England! About the fortress nothing remains unchanged, but the sun which at evening still brightens it in its weak old age with the same glory that shone over its lusty youth; the sea that rolls and dashes, as at first, against its foundation rocks; and the wild Cornish country outspread on either side of it, as desolately and as magnificently as ever.

The grandeur of the scenery at Tintagel, the romantic interest of the old British traditions connected with the castle, might well have delayed us many hours on these solitary heights; but we had other places still to visit, other and far different legends still to gossip over. Descending the cliff while the day gave us ample time to wander at our will; we strolled away inland to track the scene of a new romance as far as the waterfall called Nighton's Keive.

A walk of little more than half-a-mile brings us to the entrance of a valley, bounded on either side by high, gently-sloping hills, with a small stream running through its centre, fed by the waterfall of which we are in search. We now follow a footpath a few hundred yards, pass by a mill, and looking up the valley, see one compact mass of vegetation entirely filling it to its remotest corners, and not leaving the slightest vestige of a path, the merest patch of clear ground, visible in any direction, far or near.

It seems as if all the foliage which ought to have grown on the Cornish moorlands, had been mischievously crammed into this place, within the narrow limits of one Cornish valley. Weeds, ferns, brambles, bushes, and young trees, are flourishing together here, thickly intertwined in every possible position, in triumphant security from any invasion of bill-hook or axe. You win every step of your way through this miniature forest of vegetation, by the labour of your arms and the weight of your body. Tangled branches and thorny bushes press against you in front and behind, meet over your head, knock off your cap, flap in your face, twist about your legs, and tear your coat skirts; so obstructing you in every conceivable manner and in every conceivable direction, that they seem possessed with a living power of opposition, and commissioned by some evil genius of Fairy Mythology to prevent mortal footsteps from intruding into the valley. Whether you try a zigzag or a straight course, whether you go up or down, it is the same thing—you must squeeze, and push, and jostle your way through the crowd of bushes, just as you would through a crowd of men—or else stand still, surrounded by leaves, like " a Jack-in-the-Green," and wait for the very remote chance of somebody coming to help you out.

Forcing our road incessantly through these obstructions, for a full half-hour, and taking care to keep our only guide—the sound of the running-water—always within hearing, we came at last to a little break in the vegetation, crossed the stream at this place, and found, on the opposite side of the bank, a faintly-marked track, which might have been once a footpath. Following it, as well as we could among the branches and brambles, and now ascending steep ground, we soon heard the dash of the waterfall. But to attempt to see it, was no easy undertaking. The trees, the bushes, and the wild herbage grew here thicker than ever, stretching in perfect canopies of leaves so closely across the overhanging banks of the stream, as entirely to hide it from view. We heard the monotonous, eternal splashing of the water, close at our ears, and yet vainly tried to obtain even a glimpse of the fall. Adverse Fate led us up and down, and round and round, and backwards and forwards, amid a labyrinth of overgrown bushes which might have bewildered an Australian settler; and still the nymph of the waterfall coyly hid herself from our eyes. Our ears informed us that the invisible object of which we were in search was of very inconsiderable height; our patience was evaporating; our time was wasting away—in short, to confess the truth here, as I have confessed it elsewhere in these pages, let me acknowledge that we both concurred in a sound determination to consult our own convenience, and give up the attempt to discover Nighton's Keive!

Our wanderings, however, though useless enough

in one direction, procured us this compensating advantage in another: they led us accidentally to the exact scene of the legend which we knew to be connected with this part of the valley, and which had, indeed, first induced us to visit it.

We found ourselves standing before the damp, dismantled stone walls of a solitary cottage, placed on a plot of partially open ground, near the outskirts of the wood. Long dark herbage grew about the inside of the ruined little building; a toad was crawling where the leaves clustered thickest, on what had once been the floor of a room; in every direction corruption and decay were visibly battening on the lonesome place. Its aspect would repel rather than allure curiosity, but for the mysterious story associated with it, which gives it an attraction and an interest that are not its own.

Years and years ago, when this desolate building was a neat comfortable cottage, it was inhabited by two ladies, of whose histories, and even names, all the people of the district were perfectly ignorant. One day they were accidentally found living in their solitary abode, before any one knew that they had so much as entered it, or that they existed at all. Both appeared to be about the same age, and both were inflexibly taciturn. One was never seen without the other; if they ever left the house, they only left it to walk in the most unfrequented parts of the wood; they kept no servant, and never had a visitor; no living souls but themselves ever crossed the door of their cottage. They procured their food and other necessaries from the people in the nearest village, paying for everything they received when it was delivered, and neither asking nor answering a single unnecessary question. Their manners were gentle, but grave and sorrowful as well. The people who brought them their household supplies, felt awed and uneasy, without knowing why, in their presence; and were always relieved when they had dispatched their errand and had got well away from the cottage and the wood.

Gradually, as month by month passed on, and the mystery hanging over the solitary pair was still not cleared up, superstitious doubts spread widely through the neighbourhood. Harmless as the conduct of the ladies always appeared to be, there was something so sinister and startling about the unearthly seclusion and secrecy of their lives, that people began to feel vaguely suspicious, to whisper awful imaginary rumours about them, to gossip over old stories of ghosts and false accusations that had never been properly sifted to the end, whenever the inhabitants of the cottage were mentioned. At last they were secretly watched by the less scrupulous among the villagers, whom intense curiosity had endowed with a morbid courage and resolution. Even this proceeding led to no results whatever, but increased rather than diminished the mystery.

The expertest eavesdroppers who had listened at the door, brought away no information with them for their pains. Some declared that when the ladies held any conversation together, they spoke in so low a tone that it was impossible to distinguish a word they said. Others, of more imaginative temperament, protested, on the contrary, that their voices were perfectly audible, but that the language they talked was some mysterious or diabolical language of their own, incomprehensible to everybody but themselves. One or two expert and daring spies had even contrived to look in at them through the window, unperceived; but had seen nothing uncommon, nothing supernatural,—nothing, in short, beyond the spectacle of two ladies sitting quietly and silently by their own fireside.

So matters went on, until one day universal agitation was excited in the neighbourhood by a rumour that one of the ladies was dead. The rustic authorities immediately repaired to the cottage, accompanied by a long train of eager followers; and found that the report was true. The surviving lady was seated by her companion's bedside, weeping over a corpse. She spoke not a word; she never looked up at the villagers as they entered. Question after question was put to her without ever eliciting an answer; kind words were useless — even threats proved equally inefficient: the lady still remained weeping by the corpse, and still said nothing. Gradually her inexorable silence began to infect the visitors to the cottage. For a few moments nothing was heard in the room but the dash of the waterfall hard by, and the singing of birds in the surrounding wood. Bitterly as the lady was weeping, it was now first observed by everybody that she wept silently, that she never sobbed, never even sighed under the oppression of her grief.

People began to urge each other, superstitiously, to leave the place. It was determined that the corpse should be removed and buried; and that afterwards some new expedient should be tried to induce the survivor of the mysterious pair to abandon her inflexible silence. It was anticipated that she would have made some sign, or spoken some few words when they lifted the body from the bed on which it lay; but even this proceeding produced no visible effect. As the villagers quitted the dwelling with their dead burden, the last of them who went out left her in her solitude, still speechless, still weeping, as they had found her at first.

Days passed, and she sent no message to any one. Weeks elapsed, and the idlers who waited about the woodland paths where they knew that she was once wont to walk with her companion, never saw her, watch for her as patiently as they might. From haunting the wood, they soon got on to hovering round the cottage, and to looking in stealthily at the window. They saw her sitting on the same seat that she had always occupied, with a vacant chair opposite; her figure wasted, her face wan already with incessant weeping. It was a dismal sight to all who beheld it—a vision of affliction and solitude that sickened their hearts.

No one knew what to do; the kindest-hearted people hesitated, the hardest-hearted people dreaded to disturb her. While they were still irresolute, the end was at hand. One morning a little girl, who had looked in at the cottage window in imitation of her elders, reported, when she returned home, that she had seen the lady still sitting in her accustomed place, but that one of her hands hung strangely over the arm of the chair, and that she never moved to pick up her pocket-handkerchief, which lay on the ground beside her. At these ominous tidings, the villagers summoned their resolution, and immediately repaired to the lonesome cottage in the wood.

They knocked and called at the door—it was not opened to them. They raised the latch and entered. She still occupied her chair; her head was resting on one of her hands; the other hung down, as the little girl had told them. The handkerchief, too, was on the ground, and was wet with tears. Was she sleeping? They went round in front to look. Her eyes were wide open; her drooping hand, worn almost to mere bone, was cold to the touch as the waters of the valley-stream on a winter's day. She had died in her wonted place; died in mystery and in solitude as she had lived.

They buried her where they had buried her companion. No traces of the real history of either the one or the other have ever been discovered from that time to this.

Such is the tale that was related to us of the cottage in the valley of Nighton's Keive. It may be only imagination; but the stained roofless walls, the damp clotted herbage, and the reptiles crawling about the ruins, give the place a gloomy and disastrous look. The air, too, seems just now unusually still and heavy here—for the evening is at hand, and the vapours are rising in the wood. The shadows of the trees are deepening; the rustling music of the waterfall is growing dreary; the utter stillness of all things besides, becomes wearying to the ear. Let us pass on, and get into bright wide space again, where the down leads back to happier solitudes by the sea-shore.

We now rapidly lose sight of the trees which have hitherto so closely surrounded us, and find ourselves treading the short scanty grass of the cliff-top once more. We still advance northward, walking along rough cart-roads, and skirting the extremities of narrow gullies leading down to the sea, until we enter the picturesque village of Boscastle. Then, descending a long street of irregular houses, of all sizes, shapes, and ages, we are soon conducted to the bottom of a deep hollow. Beyond this, the bare ground rises again abruptly up to the highest point of the high cliffs which overhang the shore; and here, where the site is most elevated, and where neither cottages nor cultivation appear, we descry the ancient walls and gloomy tower of Forrabury Church.

The interior of the building still contains a part of the finely-carved rood-loft which once adorned it. Its rickety wooden pews are blackened with extreme old age, and covered with curiously-cut patterns and cyphers. The place is so dark that it is difficult to read the inscriptions on many of the mouldering monuments, fixed together without order or symmetry on the walls. Outside are some Saxon arches, oddly built of black slate-stone; and the window-mouldings are ornamented with rough carving, which at once proclaims its own antiquity. But it is in the tower that the interest attached to the church chiefly centres. Square, thick, and of no extraordinary height, it resembles in appearance most other towers in Cornwall —except in one particular, all the belfry windows are completely stopped up.

This peculiarity is to be explained simply enough: the church has never had any bells; the old tower has been mute, and useless except for ornament, since it was first built. The congregation of the district must trust to their watches and their puctuality to get to service in good time on Sundays. At Forrabury the chimes have never sounded for a marriage: the knell has never been heard for a funeral.

To know the reason of this; to discover why the church, though tower and belfry have always been waiting ready for them, has never had a peal of bells, we must seek instruction from another popular tradition, from a third legend of these legendary shores. Let us go down a little to the brink of the cliff, where the sea is rolling into a black, yawning, perpendicular pit of slate rock. The scene of our third story is the view over the waters from this place.

In ancient times, when Forrabury Church was still regarded as a building of recent date, it was a subject of sore vexation to all the people of the neighbourhood that their tower had no bells, while the inhabitants of Tintagel still possessed the famous peal that had rung for King Arthur's funeral. For some years, this superiority of the rival village was borne with composure by the people of Forrabury; but, in process of time, they lost all patience, and it was publicly determined by the rustic council, that the honour of their church should be vindicated. Money was immediately collected, and bells of magnificent tones and dimensions were forthwith ordered from the best manufactory that London could supply.

The bells were cast, blessed by high ecclesiastical authorities, and shipped for transportation to Forrabury. The voyage was one of the most prosperous that had ever been known. Fair winds and calm seas so expedited the passage of the ship, that she appeared in sight of the downs on which the church stood, many days before she had been expected. Great was the triumph of the populace on shore, as they watched her working into the bay with a steady evening breeze.

On board, however, the scene was very different. Here there was more uproar than happiness, for the captain and the pilot were at open opposition. As the ship neared the harbour, the bells of Tintagel were faintly heard across the water, ringing for the evening service. The pilot, who was a devout man, took off his hat as he heard the sound, crossed himself, and thanked God aloud for a prosperous voyage. The captain, who was a reckless, vain-glorious fellow, reviled the pilot as a fool, and impiously swore that the ship's company had only to thank his skill as a navigator, and their own strong arms and ready wills, for bringing the ship safely in sight of harbour.

The pilot, in reply, rebuked him as an infidel, and still piously continued to return thanks as before; while the captain, joined by the crew, tried to drown his voice by oaths and blasphemy. They were still shouting their loudest, when the vengeance of Heaven descended in judgment on them all.

The clouds supernaturally gathered, the wind rose to a gale in a moment. An immense sea, higher than any man had ever beheld, overwhelmed the ship; and, to the horror of the people on shore, she went down in an instant, close to land. Of all the crew, the pilot only was saved.

The bells were never recovered. They were heard tolling a muffled death-peal, as they sank with the ship; and even yet, on stormy days, while the great waves roll over them, they still ring their ghostly knell above the fiercest roaring of wind and sea.

This is the ancient story of the bells—this is why the chimes are never heard from the belfry of Forrabury Church.

Now that we have visited the scene of our third legend, what is it that keeps me and my companion still lingering on the downs? Why we are still delaying the hour of our departure long after the time which we have ourselves appointed for it?

We both know but too well. At this point we leave the coast, not to return to it again: at Forrabury we look our last on the sea from these rocky shores. With this evening, our pleasant days of strolling travel are ended. To-morrow we go direct to Launceston, and from Launceston at once to Plymouth. To-morrow the adventures of the walking tourist are ours no longer; for on that day our rambles in Cornwall will have virtually closed!

Rise, brother-traveller! We have lingered until twilight already; the seaward crags grow vast and dim around us, and the inland view narrows and darkens solemnly in the waning light. Shut up your sketch-book which you have so industriously filled, and pocket your pencils which you have worn down to stumps, even as I now shut up my dogs-eared old journal, and pocket my empty ink-bottle. One more of the few and fleeting scenes of life is fast closing, soon to leave us nothing but the remembrance that it once existed—a happy remembrance of a holiday walk in dear old England, which will always be welcome and vivid to the last, like other remembrances of home.

Come! the night is drawing round us her curtain of mist; let us strap on our trusty old friends, the knapsacks for the last time, and turn resolutely from the shore by which we have delayed too long. Come! let us once again "jog on the footpath way" as contentedly, if not quite as merrily, as ever; and, remembering how much we have seen and learnt that must surely better us both, let us, as we now lose sight of the dark, grey waters, gratefully, though sadly, speak the parting word:—

FAREWELL TO CORNWALL!

POSTSCRIPT TO

RAMBLES BEYOND RAILWAYS.

THE CRUISE OF THE TOMTIT

to

The Scilly Islands.

THE CRUISE OF THE TOMTIT.

I.

"At any other time of the year and for a shorter cruise, I should be delighted to join you. But as I prefer dying a dry death, I must decline accompanying you all the way to the Scilly Islands in a little pleasure boat of thirteen tons, just at the time of the autumnal equinox. You may meet with a gale that will blow you out of the water. You are running a risk, in my opinion, of the most senseless kind—and, if I thought my advice had any weight with you, I should say most earnestly, be warned in time, and give up the trip."—*Extract from the letter of A Prudent Friend.*

"If I were only a single man, there is nothing I should like better than to join you. But I have a wife and family, and I can't reconcile it to my conscience to risk being drowned."—*Report from the Personal Statement of a Married Friend.*

"Don't come back bottom upwards."—*Final Valedictory Blessing of a Facetious Friend.*

My messmate and I, having absolutely made up our minds to go to the Scilly Islands, received the expressions of opinion quoted above, with the supreme composure which distinguishes all resolute men. In other words, we held fast to our original determination, engaged the boat and the crew, and put to sea on our appointed day, in the teeth of the wind and of our friends' objections. But before I float the present narrative into blue water, I have certain indispensable formalities to accomplish which will keep me and my readers for a little while yet on dry land. First of all, let me introduce our boat, our crew, and ourselves.

Our boat is named the Tomtit. She is cutter-rigged. Her utmost length from stem to stern is thirty-six feet, and her greatest breadth on deck is ten feet. As her size does not admit of bulwarks, her deck, between the cabin-hatch and the stern, dips into a kind of well, with seats round three sides of it, which we call the Cockpit. Here we can stand up in rough weather without any danger of being rolled overboard; elsewhere, the sides of the vessel do not rise more than a few inches above the deck. The cabin of the Tomtit is twelve feet long, eight feet wide, and five feet six inches high. It has roomy lockers, and a snug little fireplace, and it leads into two recesses forward, which make capital storerooms for water, coals, firewood, and so forth. When I have added that the Tomtit has a bright red bottom, continued, as to colour, up her sides to a little above the watermark; and when I have further stated that she is a fast sailer, and that she proved herself on our cruise to be a capital little seaboat, I have said all that is needful at present on the subject of our yacht, and may get on to our crew and ourselves.

Our crew is composed of three brothers: Sam Dobbs, Dick Dobbs, and Bob Dobbs; all active seamen, and as worthy and hearty fellows as any man in the world could wish to sail with. My friend's name is Mr. Migott, and mine is Mr. Jollins. Thus, we are five on board altogether. As for our characters, I shall leave them to come out as they may in the course of this narrative. I am going to tell things plainly just as they happened. Smart writing, comic colouring, and graphic description, are departments of authorship at which I snap my fingers in contempt.

The port we sailed from was a famous watering-place on the western coast, called Mangerton-on-the-Mud; and our intention, as intimated at the beginning of these pages, was to go even farther than the Land's End, and to reach those last morsels of English ground called the Scilly Islands. But if the reader thinks he is now to get afloat at once, he is lamentably mistaken. One very important and interesting part of our voyage was entirely comprised in the preparations that we made for it. To this portion of the subject, therefore, I shall wholly devote myself in the first instance. On paper, or off it, neither Mr. Migott nor myself are men to be hurried.

We left London with nothing but our clothes, our wrappers, some tobacco, some French novels, and some Egyptian cigars. Everything that was to be bought for the voyage was to be procured at Bristol. Everything that could be extracted from private benevolence, was to be taken in unlimited quantities from hospitable friends living more or less in the neighbourhood of our place of embarkation. At Bristol we plunged over head and ears in naval business immediately. After ordering a ham, and a tongue, marmalade, lemons, anchovy paste, and general groceries, we set forth to the quay to equip ourselves and our vessel.

We began with charts, sailing directions, and a compass; we got on to a hammock a-piece and a flag; and we rose to a nautical climax by buying tarpaulin-coats, leggings, and sou'-westers, at a sailors' public-house. With these sea-stores, and with a noble loaf of home-made bread (the offering of private benevolence) we left Bristol to scour the friendly country beyond, in search of further contributions to the larder of the Tomtit.

The first scene of our ravages was a large country-house, surrounded by the most charming grounds. From the moment when we and our multifarious packages poured tumultuous into the hall, to the moment when we and the said packages poured out of it again into a carriage and a cart, I have no recollection, excepting meal-times and bedtime, of having been still for an instant. Escorted everywhere by two handsome, high-spirited boys, in a wild state of excitement about our voyage, we ranged the house from top to bottom, and laid hands on everything portable and eatable that we wanted in it. The inexhaustible hospitality of our hostess was proof against all the inroads that we could make on it. The priceless gift of packing perishable commodities securely in small spaces, possessed by a lady living in the house and placed perpetually at our disposal, encouraged our propensities for unlimited accumulation. We ravaged the kitchen garden and the fruit-garden; we rushed into the awful presence of the cook (with our ham and tongue from Bristol as an excuse) and ranged predatory over the lower regions. We scaled back-staircases, and tramped along remote corridors, and burst into secluded lumber-rooms, with accompaniment of shouting from the boys, and of operatic humming from Mr. Migott and myself, who happen, among other social accomplishments, to be both of us musical in a desultory way. We turned out, in these same lumber-rooms, plans of estates from their neat tin cases, and put in lemons and loaf-sugar instead. Mr. Migott pounced upon a stray telescope, and strapped it over my shoulders forthwith. The two boys found two japanned boxes, with the epaulettes and shako of an ex-military member of the family inside, which articles of martial equipment (though these are war-times, and nobody is meritorious or respectable now who does not wear a uniform) I, with my own irreverent hands, shook out on the floor; and straightway conveyed the empty cases down-stairs to be profaned by tea, sugar, Harvey's sauce, pickles, pepper, and other products of the arts of peace. In a word, and not to dwell too long on the purely piratical part of our preparations for the voyage, we doubled the number of our packages at this hospitable country house, before we left it for Mangerton-on-the-Mud, and the dangers of the sea that lay beyond.

At Mangerton we made a second piratical swoop upon another long-suffering friend, the resident doctor. We let this gentleman off, however, very easily, only lightening him of a lanthorn, and two milk-cans to hold our fresh-water. We felt strongly inclined to take his warmest cape away from him also; but Mr. Migott leaned towards the side of mercy, and Mr. Jollins was, as usual, only too ready to sacrifice himself on the altar of friendship—so the doctor kept his cape, after all.

Not so fortunate was our next victim, Mr. Purler, the Port Admiral of Mangerton-on-the-Mud, and the convivial host of the Metropolitan Inn. Wisely entering his house empty-handed, we left it with sheets, blankets, mattresses, pillows, table-cloths, napkins, knives, forks, spoons, crockery, a frying-pan, a gridiron, and a saucepan. When to these articles of domestic use were added the parcels we had brought from Bristol, the packages we had collected at the country-house, the doctor's milk-cans, the personal baggage of the two enterprising voyagers, additions to the eating and drinking department in the shape of a cold curry in a jar, a piece of spiced beef, a side of bacon, and a liberal supply of wine, spirits, and beer—nobody can be surprised to hear that we found some difficulty in making only one cart-load of our whole collection of stores. The packing process was, in fact, not accomplished till after dark. The tide was then flowing; we were to sail the next morning; and it was necessary to get everything put on board that night, while there was water enough for the Tomtit to be moored close to the jetty.

This jetty, it must be acknowledged, was nothing but a narrow stone causeway, sloping down from the land into the sea. Our cart, loaded with breakable things, was drawn up at the high end of the jetty; the Tomtit waiting to receive the contents of the cart at the low end, in the water. We had no moon, no stars, no lamp of any kind on shore; and the one small lanthorn on board the vessel just showed how dark it was, and did nothing more. Imagine the doctor, and the doctor's friend, and the doctor's two dogs, and Mr. Migott and Mr. Jollins, all huddled together in a fussy state of expectation, midway on the jetty, seeing nothing, doing nothing, and being very much in the way—and then wonder, as we wondered, at the marvellous dexterity of our three valiant sailors, who succeeded in transporting piecemeal the crockery, cookery, and general contents of the cart into the vessel, on that pitchy night, without breaking, dropping, or forgetting anything. When I hear of professional conjurors performing remarkable feats, I think of the brothers Dobbs, and the loading of the Tomtit in the darkness; and I ask myself if any landsman's mechanical legerdemain can be more extraordinary than the natural neat-handedness of a sailor?

The next morning the sky was black, the wind was blowing hard against us, and the waves were showing their white frills angrily in the offing. A double row of spectators had assembled at the jetty, to see us beat out of the bay. If they had come to see us hanged, their grim faces could not have expressed greater commiseration. Our only cheerful farewell came from the doctor and his friend and the two dogs. The remainder of the spectators evidently felt that they were having a last long stare at us, and that it would be indecent and unfeeling, under the circumstances, to look happy. Produce me a respectable inhabitant of an English country town, and I will match him, in the matter of stolid and silent staring, against any other man, civilized or savage, over the whole surface of the globe.

If we had felt any doubts of the sea-going qualities of the Tomtit, they would have been solved when we "went about," for the first time, after leaving the jetty. A livelier, stiffer, and drier little vessel of her size never was built. She jumped over the waves, as if the sea was a great play-ground, and the game for the morning, Leap-Frog. Though the wind was so high that we were obliged to lower our foresail, and to double-reef the mainsail, the only water we got on board was the spray that was blown over us from the tops of the waves. In the state of the weather, getting down Channel was out of the question. We were obliged to be contented, on this first day of our voyage, with running across to the Welsh coast, and there sheltering ourselves—amid a perfect fleet of outward-bound merchantmen driven back by the wind—in a snug roadstead, for the afternoon and the night.

This delay, which might have been disagreeable enough later in our voyage, gave us just the time we wanted for setting things to rights on board.

Our little twelve-foot cabin, it must be remembered, was bed-room, sitting-room, dining-room, store-room, and kitchen, all in one. Everything we wanted for sleeping, reading, eating, and drinking, had to be arranged in its proper place. The butter and candles, the soap and cheese, the salt and sugar, the bread and onions, the oil-bottle and the brandy-bottle, for example, had to be put in places where the motion of the vessel could not roll them together, and where, also, we could any of us find them at a moment's notice. Other things, not of the eatable sort, we gave up all idea of separating. Mr. Migott and I mingled our stock of shirts as we mingled our sympathies, our fortunes, and our flowing punch-bowl after dinner. We both of us have our faults; but incapability of adapting ourselves cheerfully to circumstances is not among them. Mr. Migott, especially, is one of those rare men who could dine politely off blubber in the company of Esquimaux, and discover the latent social advantages of his position if he was lost in the darkness of the North Pole.

After the arrangement of goods and chattels, came

dinner (the curry warmed up with a second course of fried onions)—then the slinging of our hammocks by the neat hands of the Brothers Dobbs—and then the practice of how to get into the hammocks, by Messrs. Migott and Jollins. No landsman who has not tried the experiment can form the faintest notion of the luxury of the sailor's swinging bed, or of the extraordinary difficulty of getting into it for the first time. The preliminary action is to stand with your back against the middle of your hammock, and to hold by the edge of the canvas on either side. You then duck your head down, throw your heels up, turn round on your back, and let go with your hands, all at the same moment. If you succeed in doing this, you are in the most luxurious bed that the ingenuity of man has ever invented. If you fail, you measure your length on the floor. So much for hammocks.

After learning how to get into bed, the writer of the present narrative tried his hand at the composition of whisky punch, and succeeded in imparting satisfaction to his intemperate fellow-creatures. When the punch and the pipes accompanying it had come to an end, a pilot-boat anchored alongside of us for the night. Once embarked on our own element, we old sea-dogs are, after all, a polite race of men. We asked the pilot where he had come from—and he asked us. We asked the pilot where he was bound to, to-morrow morning—and he asked us. We asked the pilot whether he would like a drop of rum—and the pilot, to encourage us, said Yes. After that, there was a little pause; and then the pilot asked us, whether we would come on board his boat—and we, to encourage the pilot, said Yes, and did go, and came back, and asked the pilot whether he would come on board our boat—and he said Yes, and did come on board, and drank another drop of rum. Thus in the practice of the social virtues did we while away the hours—six jolly tars in a twelve-foot cabin—till it was past eleven o'clock, and time, as we say at sea, to tumble in, or tumble out, as the case may be, when a jolly tar wants practice in the art of getting into his hammock.

So began and ended our first day afloat.

II.

THE wind blew itself out in the night. As the morning got on, it fell almost to a calm; and the merchantmen about us began weighing anchor, to drop down Channel with the tide. The Tomtit, it is unnecessary to say, scorned to be left behind, and hoisted her sails with the best of them. Favoured by the lightness of the wind, we sailed past every vessel proceeding in our direction. Barques, brigs, and schooners, French luggers and Dutch galliots, we showed our stern to all of them; and when the weather cleared, and the breeze freshened towards the afternoon, the little Tomtit was heading the whole fleet.

In the evening we brought up close to the high coast of Somersetshire, to wait for the tide. Weighed again, at ten at night, and sailed for Ilfracombe. Got becalmed towards morning, but managed to reach our port at ten, with the help of the sweeps, or long oars. Went ashore for more bread, beer, and fresh water; feeling so nautical by this time, that the earth was difficult to walk upon; and all the people we had dealings with presented themselves to us in the guise of unmitigated land-sharks. O, my dear eyes! what a relief it was to Mr. Migott and myself to find ourselves in our floating castle, boxing the compass, dancing the hornpipe, and splicing the mainbrace freely in our ocean-home.

About noon we sailed for Clovelly. Our smooth

passage across the magnificent Bay of Bideford is the recollection of our happy voyage which I find myself looking back on most admiringly while I now write.

No cloud was in the sky. Far away, on the left, sloped inward the winding shore; so clear, so fresh, so divinely tender in its blue and purple hues, that it was the most inexhaustible of luxuries only to look at it. Over the watery horizon, to the right, the autumn sun hung grandly, with the fire-path below heaving on a sea of lustrous blue. Flocks of wild birds at rest, floated chirping on the water all around. The fragrant steady breeze was just enough to fill our sails. On and on we went, with the bubbling sea-song at our bows to soothe us; on and on, till the blue lustre of the ocean grew darker, till the sun sank redly towards the far water-line, till the sacred evening stillness crept over the sweet air, and hushed it with a foretaste of the coming night.

What sight of mystery and enchantment rises before us now? Steep, solemn cliffs, bare in some places—where the dark-red rock has been rent away, and the winding chasms open grimly to the view —but clothed for the most part with trees, which soften their summits into the sky, and sweep all down them, in glorious masses of wood, to the very water's edge. Climbing from the beach, up the precipitous face of the cliff, a little fishing village coyly shows itself. The small white cottages rise one above another; now perching on a bit of rock, now peeping out of a clump of trees: sometimes two or three together; sometimes one standing alone; here, placed sideways to the sea, there, fronting it,—but rising always one over the other, as if, instead of being founded on the earth, they were hung from the trees on the top of the cliff. Over all this lovely scene the evening shadows are stealing. The last rays of the sun just tinge the quiet water, and touch the white walls of the cottages. From out at sea comes the sound of a horn—blown from the nearest fishing-vessel, as a signal to the rest to follow her to shore. From the land, the voices of children at play, and the still fall of the small waves on the beach, are the only audible sounds. This is Clovelly. If we had travelled a thousand miles to see it, we should have said that our journey had not been taken in vain.

On getting to shore, we found the one street of Clovelly nothing but a succession of irregular steps, from the beginning at the beach, to the end half way up the cliffs. It was like climbing to the top of an old castle, instead of walking through a village.

When we reached the summit of the cliff, the hour was too advanced to hope for seeing much of the country. We strayed away, however, to look for the church, and found ourselves, at twilight, near some ghastly deserted out-houses, approached by a half-ruinous gateway, and a damp dark avenue of trees. The church was near, but shut off from us by ivy-grown walls. No living creature appeared; not even a dog barked at us. We were surrounded by silence, solitude, darkness, and desolation; and it struck us both forcibly, that the best thing we could do was to give up the church, and get back to humanity with all convenient speed.

The descent of the High Street of Clovelly, at night, turned out to be a matter of more difficulty than we had anticipated. There was no such thing as a lamp in the whole village; and we had to grope our way in the darkness down steps of irregular sizes and heights, paved with slippery pebbles, and ornamented with nothing in the shape of a bannister, even at the most dangerous places. Half-way down, my friend and I had an argument in the dark—standing with our noses against a wall, and with nothing visible on either side—as to which way we should turn next. I guessed to the left, and he guessed to the right; and I, being the more obstinate of the two, we ended in following my route, and at last stumbled our way down to the pier. Looking at the place the next morning, we found that the steps to the right led through a bit of cottage-garden to a snug little precipice, over which inquisitive tourists might fall quietly, without let or hindrance. Talk of the perils of the deep! what are they in comparison with the perils of the shore?

The adventures of the night were not exhausted, so far as I was concerned, even when we got back to our vessel.

I have already informed the reader that the cabin of the Tomtit was twelve feet long by eight feet wide—a snug apartment, but scarcely large enough, as it struck me, for five men to sleep in comfortably. Nevertheless, the experiment was to be tried in Clovelly harbour. I bargained, at the outset, for one thing—that the cabin hatch should be kept raised at least a foot all night. This ventilatory condition being complied with, I tumbled into my hammock; Mr. Migott rolled into his; and Sam Dobbs, Dick Dobbs, and Bob Dobbs cast themselves down promiscuously on the floor and the lockers under us. Out went the lights; and off went my friend and the

Brothers Dobbs into the most intolerable concert of snoring that it is possible to imagine.

No alternative was left for my unfortunate self but to lie awake listening, and studying the character of the snore in each of the four sleeping individuals. The snore of Mr. Migott I found to be superior to the rest in point of amiability, softness, and regularity—it was a kind of oily, long-sustained purr, amusing and not unmusical for the first five minutes. Next in point of merit to Mr. Migott, came Bob Dobbs. His note was several octaves lower than my friend's, and his tone was a grunt—but I will do him justice; I will not scruple to admit that the sounds he produced were regular as clockwork. Very inferior was the performance of Sam Dobbs, who, as owner of the boat, ought, I think, to have set a good example. If an idle carpenter planed a board very quickly at one time, and very slowly at another, and if he groaned at intervals over his work, he would produce the best imitation of Sam Dobbs's style of snoring that I can think of. Last, and worst of all, came Dick Dobbs, who was afflicted with a cold, and whose snore consisted of a succession of loud chokes, gasps, and puffs, all contending together, as it appeared to me, which should suffocate him soonest. There I lay, wide

T

awake, suffering under the awful nose-chorus which I have attempted to describe, for nearly an hour. It was a dark night: there was no wind, and very little air. Horrible doubts about the sufficiency of our ventilation began to beset me. Reminiscences of early reading on the subject of the Black Hole at Calcutta came back vividly to my memory. I thought of the twelve feet by eight, in which we were all huddled together—terror and indignation overpowered me—and I roared for a light, before the cabin of the Tomtit became too mephitic for flame of any kind to exist in it. Uprose they then my Merry Merry Men, bewildered and grumbling, to grope for the match-box. It was found, the lantern was lit, the face of Mr. Migott appeared serenely over the side of his hammock, and the voice of Mr. Migott sweetly and sleepily inquired what was the matter?

"Matter! The Black Hole at Calcutta is the matter. Poisonous, gaseous exhalation is the matter! Outrageous, ungentlemanly snoring is the matter! give me my bedding, and my drop of brandy, and my pipe, and let me go on deck. Let me be a Chaldean shepherd, and contemplate the stars. Let me be the careful watch who patrols the deck, and guards the ship from foes and wreck. Let me be anything but the companion of men who snore like the famous Furies in the old Greek play." While I am venting my indignation, and collecting my bedding, the smiling and sleepy face of Mr. Miggott disappears slowly from the side of the hammock—and before I am on deck, I hear the oily purr once more, just as amiable, soft, and regular as ever.

What a relief it was to have the sky to look up at, the fresh night air to breathe, the quiet murmur of the sea to listen to! I rolled myself up in my blankets; and, for aught I know to the contrary, was soon snoring on deck as industriously as my companions were snoring below.

The first sounds that woke me in the morning were produced by the tongues of the natives of Clovelly, assembled on the pier, staring down on me in my nest of blankets, and shouting to each other incessantly. I assumed that they were making fun of the interesting stranger stretched in repose on the deck of the Tomtit; but I could not understand one word of the Devonshire language in which they spoke. Whatever they said of me, I forgive them, however, in consideration of their cream and fresh herrings. Our breakfast on the cabin-hatch in Clovelly harbour, after a dip in the sea, is a remembrance of gustatory bliss which I gratefully cherish. When we had reduced the herrings to skeletons, and the cream-pot to a whited sepulchre of emptiness, we slipped from our moorings, and sailed away from the lovely little village with sincere regret. By noon we were off Hartland Point.

We had now arrived at the important part of our voyage—the part at which it was necessary to decide, once for all, on our future destination. Mr. Migott and I took counsel together solemnly, unrolled the charts, and then astonished our trusty crew by announcing that the end of the voyage was to be the Scilly Islands. Up to this time the Brothers Dobbs had been inclined to laugh at the notion of getting so far in so small a boat. But they began to look grave now, and to hint at cautious objections. The weather was certainly beautiful; but then the wind was dead against us. Our little vessel was stiff and sturdy enough for any service, but nobody on board knew the strange waters into which we were going—and, as for the charts, could any one of us study them with a proper knowledge of the science of navigation? Would it not be better to take a little cruise to Lundy Island, away there on the starboard bow? And another little cruise about the Welsh coast, where the Dobbses had been before? To these cautious questions, we replied by rash and peremptory negatives; and the Brothers, thereupon, abandoned their view of the case, and accepted ours with great resignation.

For the Scilly Islands, therefore, we now shaped our course, alternately standing out to sea, and running in for the land, so as to get down ultimately to the Land's End, against the wind, in a series of long zig-zags, now in a westerly and now in an easterly direction. Our first tack from Hartland Point was a sail of six hours out to sea. At sunset, the little Tomtit had lost sight of land for the first time since she was launched, and was rising and falling gently on the long swells of the Atlantic. It was a deliciously calm, clear evening, with every promise of the fine weather lasting. The spirits of the Brothers Dobbs, when they found themselves at last in the blue water, rose amazingly.

"Only give us decent weather, sir," said Bob Dobbs, cheerfully smacking the tiller of the Tomtit; "and we'll find our way to Scilly somehow, in spite of the wind."

How we found our way, remains to be seen.

III.

We were now fairly at sea, keeping a regular watch on deck at night, and never running nearer the Cornish coast than was necessary to enable us to compare the great headlands with the marks on our chart. Under present circumstances, no more than three of us could sleep in the cabin at one time—the combined powers of the snoring party were thus weakened, and the ventilation below could be preserved in a satisfactory state. Instead of chronicling our slow zig-zag progress to the Land's End—which is unlikely to interest anybody not familiar with Cornish names and nautical phrases—I will try to describe the manner in which we passed the day on board the Tomtit, now that we were away from land events and amusements. If there was to be any such thing as an alloy of dulness in our cruise, this was assuredly the part of it in which Time and the Hour were likely to run slowest through the day.

In the first place, let me record with just pride, that we have solved the difficult problem of a pure republic in our modest little craft. No man in particular among us is master—no man in particular is servant. The man who can do at the right time, and in the best way, the thing that is most wanted, is always the hero of the situation among us. When Dick Dobbs is frying the onions for dinner, he is the person most respected in the ship, and Mr. Migott and myself are his faithful and expectant subjects. When grog is to be made, or sauces are to be prepared, Mr. Jollins becomes in his turn the monarch of all he surveys. When musical entertainments are in progress, Mr. Migott is vocal king, and sole conductor of band and chorus. When nautical talk and sea-stories rule the hour, Bob Dobbs, who has voyaged in various merchantmen all over the world, and is every inch of him a thorough sailor, becomes the best man of the company. When any affairs connected with the internal management of the vessel are under consideration, Sam Dobbs is Chairman of the Committee in the cockpit. So we sail along; and such is the perfect constitution of society at which we mariners of England have been able to arrive.

Our freedom extends to the smallest details. We have no stated hours, and we are well a-head of all rules and regulations. We have no breakfast hour, no dinner hour, no time for rising or for going to bed. We have no particular eatables at particular meals. We don't know the day of the month, or the day of the week; and never look at our watches, except when we wind them up. Our voice is frequently the voice of the sluggard; but we never complain, because nobody ever wakes us too soon, or thinks of interfering with our slumbering again. We wear each other's coats, smoke each other's pipes, poach on each other's victuals. We are a happy, dawdling, undisciplined, slovenly lot. We have no principles, no respectability, no business, no stake in the country, no knowledge of Mrs. Grundy. We are a parcel of Lotos-Eaters; and we know nothing, except that we are poking our way along anyhow to the Scilly Islands in the Tomtit.

We rise when we have had sleep enough—any time you like between seven and ten. If I happen to be on deck first, I begin by hearing the news of the weather and the wind, from Sam, Dick, or Bob at the helm. Soon the face of Mr. Migott, rosy with recent snoring, rises from the cabin, and his body follows it slowly, clad in the blue Jersey frock which he persists in wearing night and day—in the heat of noon as in the cool of evening. He cannot be prevailed upon to give any reason for his violent attachment to this garment—only wagging his head and smiling mysteriously when we ask why, sleeping or waking, he never parts with it. Well, being up, the next thing is to make the toilette. We keep our fresh water, for minor ablutions, in an old wine cask from Bristol. The colour of the liquid is a tawny yellow: it is, in fact, weak sherry and water. For the major ablutions, we have the ship's bucket and the sea, and a good stock of rough towels to finish with. The next thing is breakfast on deck. When we can catch fish (which is very seldom, though we are well provided with lines and bait) we fall upon the spoil immediately. At other times we range through our sea stores, eating anything we like, cooked anyhow we like. After breakfast we have two words to say to our box of peaches, nectarines, and grapes, from the hospitable country-house. Then the bedding is brought up to air; the deck is cleaned; the breakfast things are taken away; the pipes, cigars, and French novels are produced from the cabin; Mr. Migott coils himself up in a corner of the cockpit, and I perch upon the taffrail; and the studies of the morning begin. They end invariably in small-talk, beer, and sleep. So the time slips away cosily till it is necessary to think about dinner.

Now, all is activity on board the Tomtit. Except the man at the helm, every one is occupied with preparations for the banquet of the day. The potatoes, onions, and celery, form one department; the fire and solid cookery another; the washing of plates and dishes, knives and forks, a third; the laying of the cloth on deck a fourth; the concoction of sauces and production of bottles from the cellar a fifth. No man has any particular department assigned to him: the most active republican of the community, for the time being, plunges into the most active work, and the others follow as they please.

The exercise we get is principally at this period of the day, and consists in incessant dropping down from the deck to the cabin, and incessant scrambling up from the cabin to the deck. The dinner is a long business; but what do we care for that? We have no appointments to keep, no visitors to interrupt us, and nothing in the world to do but to tickle our palates, wet our whistles, and amuse ourselves in any way we please. Dinner at last over, it is superfluous to say, that the pipes become visible again, and that the taking of forty winks is only a prohibited operation on the part of the man at the helm.

As for tea-time, it is entirely regulated by the wants and wakefulness of Mr. Migott, who, since the death of Dr. Johnson, is the most desperate drinker of tea in all England. When the cups and saucers are cleared away, a conversazione is held in the cockpit. Sam Dobbs is the best listener of the company; Dick Dobbs, who has been a yachtsman, is the jester; Bob Dobbs, the merchant sailor, is the teller of adventures; and my friend and I keep the ball going smartly in all sorts of ways, till it gets dark, and a great drought falls upon the members of the conversazione. Then, if the mermaids are anywhere near us, they may smell the fragrant fumes which tell of sacrifice to Bacchus, and may hear, shortly afterwards, the muse of song invoked by cheerful topers. Thus the dark hours roll on jovial till the soft influences of sleep descend upon the tuneful choir, and the cabin receives its lodgers for the night.

This is the general rule of life on board the Tomtit. Exceptional incidents of all kinds—saving sea-sickness, to which nobody on board is liable—are never wanting to vary existence pleasantly from day to day. Sometimes Mr. Migott gets on from taking a nap to having a dream, and records the fact by a screech of terror, which rings through the vessel and wakes the sleeper himself, who always asks, " What's that, eh?"—never believes that the screech has not come from somebody else—never knows what he has been dreaming of—and never fails to go to sleep again before the rest of the ship's company have half done expostulating with him.

Sometimes a little interesting indigestion appears among us, by way of change. Dick Dobbs, for example (who is as bilious as an Indian nabob), is seen to turn yellow at the helm, and to steer with a glazed eye; is asked what is the matter; replies that he has "the boil terrible bad on his stomach;" is instantly treated by Jollins (M.D.) as follows:—Two teaspoonfuls of essence of ginger, two dessert-spoonfuls of brown brandy, two table spoonfuls of strong tea. Pour down patient's throat very hot, and smack his back smartly to promote the operation of the draught. What follows? The cure of Dick. How simple is medicine, when reduced to its first principles!

Another source of amusement is provided by the ships we meet with.

Whenever we get near enough, we hail the largest merchant-men in the most peremptory manner, as coolly as if we had three decks under us and an admiral on board. The large ships, for the most part paralysed by our audacity, reply meekly. Sometimes we meet with a foreigner, and get answered by inarticulate yelling or disrespectful grins. But this is a rare case; the general rule is, that we maintain our dignity unimpaired all down the Channel. Then, again, when no ships are near, there is the constant excitement of consulting our charts and wondering where we are. Every man of us has a different theory on this subject every time he looks at the chart; but no man rudely thrusts his theory on another, or aspires to govern the ideas of the rest in virtue of his superior obstinacy in backing his own opinion. Did I not assert a little while since that we were a pure republic? And is not this another and a striking proof of it?

In such pursuits and diversions as I have endeavoured to describe, the time passes quickly, happily, and adventurously, until we ultimately succeed, at four in the morning on the sixth day of our cruise, in discovering the light of the Longship's Lighthouse, which we know to be situated off the Land's End. We are now only some seven and-twenty miles from the Scilly Islands, and the discovery of the lighthouse enables us to set our course by the compass cleverly enough. The wind which has thus far always remained against us, falls, on the afternoon of this sixth day, to a dead calm, but springs up again in another and a favourable quarter at eleven o'clock at night. By daybreak we are all on the watch for the Scilly Islands. Not a sign of them. The sun rises; it is a magnificent morning; the favourable breeze still holds; we have been bowling along before it since eleven the previous night; and ought to have sighted the islands long since. But we sight nothing: no land is visible anywhere all round the horizon.

Where are we? Have we overshot Scilly?—and is the next land we are likely to see Ushant or Finisterre? Nobody knows. The faces of the Brothers Dobbs darken; and they recal to each other how they deprecated from the first this rash venturing into unknown waters. We hail two ships piteously, to ask our way. The two ships can't tell us. We unroll the charts, and differ in opinion over them more remarkably than ever. The Dobbses grimly opine that it is no use looking at charts, when we have not got a pair of parallels to measure by, and are all ignorant of the scientific parts of navigation. Mr. Migott and I manfully cheer the drooping spirits of the crew with Guinness's stout, and put a smiling face upon it. But in our innermost hearts, we think of Columbus, and feel for him.

The last resource is to post a man at the masthead (if so lofty an expression may be allowed in reference to so little a vessel as the Tomtit), to keep a look-out. Up the rigging swarms Dick the Bilious, in the lowest spirits—strains his eyes over the waters, and suddenly hails the gaping deck with a joyous shout. The runaway islands are caught at last—he sees them a-head of us—he has no objection to make to the course we are steering—nothing particular to say but "Crack on!"—and nothing in the world to do but slide down the rigging again. Contentment beams once more on the faces of Sam, Dick, and Bob. Mr. Migott and I say nothing; but we look at each other with a smile of triumph. We remember the injurious doubts of the crew when the charts were last unrolled—and think of Columbus again, and feel for him more than ever.

Soon the islands are visible from the deck, and by noon we have run in as near them as we dare without local guidance. They are low-lying, and picturesque in an artistic point of view; but treacherous-looking and full of peril to the wary nautical eye. Horrible jagged rocks, and sinister swirlings and foamings of the sea, seem to forbid the approach to them. The Tomtit is hove to—our ensign is run up half-mast high—and we fire our double-barrelled gun fiercely for a pilot.

The pilot arrives in a long, serviceable-looking boat, with a wild, handsome, dark-haired son, and a silent, solemn old man for his crew. He himself is lean, wrinkled, hungry-looking; his eyes are restless with excitement, and his tongue overwhelms us with a torrent of words, spoken in a strange accent, but singularly free from provincialisms and bad grammar. He informs us that we must have been set to the northward in the night by a current, and goes on to acquaint us with so many other things, with such a fidgety sparkling of the eyes and such a ceaseless patter of the tongue, that he fairly drives me to the fore part of the vessel out of his way. Smoothly we glide along, parallel with the jagged rocks and the swirling eddies, till we come to a channel between two islands; and, sailing through that, make for a sandy isthmus, where we see some houses and a little harbour. This is Hugh Town, the chief place in St. Mary's, which is the largest island of the Scilly group. We jump ashore in high glee, feeling that we have succeeded in carrying out the purpose of our voyage in defiance of the prognostications of all our prudent friends. At sea or on shore, how sweet is triumph, even in the smallest things!

Bating the one fact of the wind having blown from an unfavourable quarter, unvarying good fortune had, thus far, accompanied our cruise, and our luck did not desert us when we got on shore at St. Mary's. We went, happily for our own comfort, to the hotel kept by the master of the packet plying between Hugh Town and Penzance. By our landlord and his cordial wife and family we were received with such kindness and treated with such care, that we felt really and truly at home before we had been half an hour in the house. And, by way of farther familiarizing us with Scilly at first sight, who should the resident medical man turn out to be but a gentleman whom I knew. These were certainly fortunate auspices under which to begin our short sojourn in one of the remotest and wildest places in the Queen's dominions.

IV.

The Scilly Islands seem, at a rough glance, to form a great irregular circle, enclosing a kind of lagoon of sea, communicating by various channels with the main ocean all around.

The circumference of the largest of the group is, as we heard, not more than thirteen miles. Five of the islands are inhabited; the rest may be generally described as masses of rock, wonderfully varied in shape and size. Inland, in the larger islands, the earth, where it is not planted or sown, is covered with heather and with the most beautiful ferns. Potatoes used to be the main product of Scilly; but the disease has appeared lately in the island crops, and the potatoes have suffered so severely that when we filled our sack for the return voyage, we were obliged to allow for two-thirds of our supply proving unfit for use. The views inland are chiefly remarkable as natural panoramas of land and sea—the two always presenting themselves intermixed in the loveliest varieties of form and colour. On the coast, the granite rocks, though not notably high, take the most wildly and magnificently picturesque shapes. They are rent into the strangest chasms and piled up in the grandest confusion; and they look down, every here and there, on the loveliest little sandy bays, where the sea, in calm weather, is as tenderly blue and as limpid in its clearness as the Mediterranean itself. The softness and purity of the climate may be imagined, when I state that in the winter none of the freshwater pools are strongly enough frozen to bear being skated on. The balmy sea air blows over each little island as freely as it might blow over the deck of a ship.

The people have the same great merit which I had previously observed among their Cornish neighbours —the merit of good manners. We two strangers were so little stared at as we walked about, that it was almost like being on the Continent. The pilot who had taken us into Hugh Town harbour we found to be a fair specimen, as regarded his excessive talkativeness and the purity of his English, of the islanders generally. The longest tellers of very long stories, so far as my experience goes, are to be found in Scilly. Ask the people the commonest question, and their answer generally exhausts the whole subject before you can say another word. Their anxiety, whenever we had occasion to inquire our way, to guard us from the remotest chance of missing it, and the honest pride with which they told us all about local sights and marvels, formed a very pleasant trait in the general character. Wherever we went, we found the natural kindness and natural hospitality of the people always ready to welcome us.

Strangely enough, in this softest and healthiest of climates consumption is a prevalent disease. If I may venture on an opinion, after a very short observation of the habits of the people, I should say that distrust of fresh air and unwillingness to take exercise were the chief causes of consumptive maladies among the islanders. I longed to break windows in the main street of Hugh Town as I never longed to break them anywhere else. One lovely afternoon I went out for the purpose of seeing how many of the inhabitants of the place had a notion of airing their bedrooms. I found two houses with open windows— all the rest were fast closed from top to bottom, as if a pestilence were abroad instead of the softest, purest sea-breeze that ever blew. Then, again, as to walking, the people ask you seriously when you inquire your way on foot, whether you are aware that the destination you want to arrive at is three miles off! As for a pedestrian excursion round the largest island—a circuit of thirteen miles—when we talked of performing that feat in the hearing of a respectable inhabitant he laughed at the idea as incredulously as if we had proposed a swimming match to the Cornish coast. When people will not give themselves the first great chance of breathing healthily and freely as often as they can, who can wonder that consumption should be common among them?

In addition to our other pieces of good fortune, we were enabled to profit by a very kind invitation from the gentleman to whom the islands belong, to stay with him at his house, built on the site of an ancient abbey, and surrounded by gardens of the most exquisite beauty.

To the firm and benevolent rule of the present proprietor of Scilly, the islanders are indebted for the prosperity which they now enjoy. It was not the least pleasant part of a very delightful visit, to observe for ourselves, under our host's guidance, all that he had done, and was doing, for the welfare and the happiness of the people committed to his charge. From what we had heard, and from what we had previously observed for ourselves, we had formed the most agreeable impressions of the social condition of the islanders; and we now found the best of these impressions more than confirmed. When the present proprietor first came among his tenantry he found them living miserably and ignorantly. He has succoured, reformed, and taught them; and there is now, probably, no place in England where the direr hardships of poverty are so little known as in the Scilly Islands.

I might write more particularly on this topic; but I am unwilling to run the risk of saying more on the subject of these good deeds than the good-doer himself would sanction. And besides, I must remember that the object of this narrative is to record a holiday-cruise, and not to enter into details on the subject of Scilly; details which have already been put into print by previous travellers. Let me only add then, that our sojourn in the islands terminated with the close of our stay in the house of our kind entertainer. It had been blowing a gale of wind for two days before our departure; and we put to sea with a doubled-reefed mainsail, and with more doubts than we liked to confess to each other, about the prospects of the return voyage.

However, lucky we had been hitherto, and lucky we were to continue to the end. Before we had been long at sea, the wind began to get capricious; then to diminish almost to a calm; then, towards evening, to blow again, steadily and strongly, from the very quarter of all others most favourable to our return voyage. "If this holds," was the sentiment of the Brothers Dobbs, as we were making things snug for the night, "we shall be back again at Mangerton before we have had time to get half through our victuals and drink."

The wind did hold, and more than hold: and the Tomtit flew, in consequence, as if she was going to give up the sea altogether, and take to the sky for a change. Our homeward run was the most perfect contrast to our outward voyage. No tacking, no need to study the charts, no laggard luxurious dining on the cabin hatch. It was too rough for anything but picnicking in the cockpit, jammed into a corner, with our plates on our knees. I had to make the grog with one hand, and clutch at the nearest rope with the other—Mr. Migott holding the bowl while I mixed, and the man at the helm holding Mr. Migott. As for reading, it was hopeless to try it; for there was breeze enough to blow the leaves out of the book—and singing was not to be so much as thought of; for the moment you opened your mouth the wind rushed in, and snatched away the song immediately. The nearer we got to Mangerton the faster we flew. My last recollection of the sea, dates at the ghostly time of midnight. The wind had been increasing and increasing, since sunset, till it contemptuously blew out our fire in the cabin, as if the stove with its artful revolving chimney had been nothing but a farthing rushlight. When I climbed on deck, we were already in the Bristol Channel.

That last view at sea was the grandest view of the voyage. Ragged black clouds were flying like spectres all over the sky; the moonlight streaming fitful behind them. One great ship, shadowy and mysterious, was pitching heavily towards us from the land. Backward out at sea, streamed the red gleam from the lighthouse on Lundy Island; and marching after us magnificently, to the music of the howling wind, came the great rollers from the Atlantic, rushing in between Hartland Point and Lundy, turning over and over in long black hills of water, with the seething spray at their tops sparkling in the moonshine. It was a fine breathless sensation to feel our sturdy little vessel tearing along through this heavy sea—jumping stern up, as the great waves caught her—dashing the water gaily from her bows, at the return dip—and holding on her way as bravely and surely as the largest yacht that ever was built. After a long look at the sublime view around us, my friend and I went below again; and in spite of the noise of wind and sea, managed to fall asleep. The next event was a call from deck at half-past six in the morning, informing us that we were entering Mangerton Bay. By seven o'clock we were alongside the jetty again, after a run of only forty-three hours from the Scilly Islands.

Thus our cruise ended; and thus we falsified the predictions of our prudent friends, and came back with our right side uppermost. "Here's luck to you, gentlemen!"—was the toast which our honest sailor-brothers proposed, when we met together later in the day, and pledged each other in a parting cup. "Here's luck," we answered, on our side—"luck to the Brothers Dobbs; and thanks besides for hearty companionship and faithful service." And here, in the last glass with one cheer more,—here's luck to the vessel that carried us, our lively little Tomtit!

Tiny home of joyous days, may thy sea-fortunes be happy, and thy trim sails be set prosperously for many a year still, to the favouring breeze!

With those good wishes, our holiday trip closed at the time—as the record of it closes here. With those last words, the book is shut up; the reader is released; and the writer drops his pen.

THE END.

www.ingramcontent.com/pod-product-compliance
Lightning Source LLC
Chambersburg PA
CBHW022053230426
43672CB00008B/1162